Concrete Repair to EN 1504

Diagnosis, Design, Principles and Practice

Concrete Repair to EN 1504

Diagnosis, Design, Principles and Practice

Michael Raupach
Till Büttner

CRC Press
Taylor & Francis Group
Boca Raton London New York

CRC Press is an imprint of the
Taylor & Francis Group, an **informa** business

A SPON PRESS BOOK

CRC Press
Taylor & Francis Group
6000 Broken Sound Parkway NW, Suite 300
Boca Raton, FL 33487-2742

First issued in paperback 2019

ISBN-13: 978-1-4665-5746-8 (hbk)
ISBN-13: 978-0-367-86714-0 (pbk)

Visit the Taylor & Francis Web site at
http://www.taylorandfrancis.com

and the CRC Press Web site at
http://www.crcpress.com

Contents

Preface

CONCRETE REPAIR: DIAGNOSIS, DESIGN, PRINCIPLES AND PRACTICE BASED ON EUROPEAN STANDARDS

Concrete repair is a complex task requiring special knowledge from all the people involved: owners, operators, building authorities, designers, executing companies, product manufacturers and suppliers, building control authorities, and even experts. This special knowledge covers technical building regulations and standards, building materials, deterioration mechanisms, diagnosis, load-bearing capacity and safety, repair principles and methods, repair materials, execution of repair works, quality control, inspections, and monitoring, as well as maintenance and building management systems.

In recent years the repair of concrete structures has become a worldwide activity. However, the diagnosis, design, selection of products, and execution of repair works need to be adjusted to the individual condition of the buildings, thus requiring specialist knowledge among all the people involved. There are several reasons for corrosion of the concrete or the reinforcement. Only if they are clearly identified by a systematic assessment of the structure can repair measures be designed successfully. Furthermore, it is essential to estimate the further development of the condition of the structure to select the optimal time for protection or repair measures.

This book is based on the new European Standard EN 1504, which provides a systematic approach from the principles and methods of repair and protection of concrete structures through to the selection of suitable products and quality control. This approach provides an optimal basis for understanding and designing repair work. The systematics of principles and methods are new, and therefore are described in detail, explained in schematic drawings, and summarised in the tables. The details of the product standards are mainly confined to the product supplier, and therefore are not discussed in detail. However, the theoretical background of corrosion and diagnosis methods, which are not covered by the EN 1504 series but are very important for successful repair work, are described using the expertise of the authors.

This comprehensive book covers both theory and practice and contains all the relevant information needed by owners, designers, companies, and interested engineers, as well as students of civil engineering, enabling them to understand the relevant background and practical aspects of concrete repair. Finally, it should be noted that this book reflects only the perspective of the authors. For certain points of EN 1504 there may be different opinions of their meaning and how they should be handled. This book is not intended to replace engineering work by strict rules, but as a helpful tool to understand the theoretical background of options, principles, and methods for protection and repair of concrete structures.

About the authors

Michael Raupach earned a PhD from the Faculty of Civil Engineering of RWTH Aachen University, Germany, on *Chloride-Induced Corrosion of Steel in Concrete*. From 1993 to 2000 he worked as a consulting engineer in the field of repair and restoration of buildings and was business manager for the production and distribution of corrosion monitoring systems. Since 2000 he has been chair of the Institute for Building Materials Research of RWTH Aachen University, known as ibac, in the fields of corrosion, repair, and conservation of buildings. He is actively engaged in the diagnosis and design with the consulting engineers Raupach Bruns Wolff. Regarding the European Standard EN 1504, which is the basis  of this book, he has been active as the German delegate since 2002. As a member of the European Federation of Corrosion, EFC, he chairs the working party 11 on corrosion of steel in concrete.

Furthermore, he has published more than 500 papers and made several contributions to books on corrosion, repair, and the maintenance of buildings.

Till Büttner, PhD, was a research assistant at the Institute for Building Materials Research of RWTH Aachen University (ibac) and head of the working group "conservation and repair" for nearly eight years before leaving the university. He is now working for an internationally known construction company as senior site supervisor. His daily work is in the field of concrete repair and maintenance, especially of concrete and steel structures, such as highway bridges. During his time at ibac he wrote his PhD thesis on the lifetime prediction of polymer-modified glass reinforcement of concrete and was active in the field of on-site diagnosis, the evaluation of the status of concrete structures as well as developing concepts for the repair and maintenance of concrete structures with convential and innovative products such as textile-reinforced concrete.

Chapter 1

Introduction

Repair and restoration of concrete structures has become an important market. For example, the costs of the maintenance for bridges are higher than $1 billion every year. The total costs for maintenance of all types of buildings are assumed to be higher than $20 billion every year. A significant part of these costs is spent on repair and protection of concrete structures. On the contrary to the amount of work spent on the building of new structures, the market of repair and protection has grown considerably as the age of the existing infrastructure has increased. The extensive development of new methods and materials for the repair and protection of concrete structures has led to the need for standards for such works.

Figure 1.1 shows the structure of European Standard EN 1504, consisting of 10 main standards and 61 standards for test methods. As shown in Table 1.1 the series of standards is focused on Parts 2 to 7, which are the basis for CE marking of the different products and systems to be used for protection and repair of concrete structures. However, Part 9 of EN 1504 is also important, describing the principles for the use of the products, which will be explained in the following sections. Part 1 gives definitions, Part 8 regulates the quality control of the products, and Part 10 gives a general guideline for site application and quality control of the works. To allow CE marking of the products, 61 standards describing test methods for the different properties of the products had to be prepared. These standards ensure that testing of the products will be according to the same standards for all products for protection and repair of concrete structures used in Europe.

The phases of repair projects follow a logical sequence, which is dominated by engineering aspects. Figure 1.2 gives a general scheme according to Figure B1 within Annex B of EN 1504-9. This figure shows the well-known elements as assessment, planning, design, and quality control. However, the systematics of general planning is quite new compared to the existing recommendations and standards. It consists of a hierarchy of levels, namely, options, principles, and methods, which are described in more detail in Chapter 6 of this book. The product standards EN 1504-2 to –7, quality control according to EN 1504-8 and test methods of the products, are not the focus of this book.

As already mentioned, the rules for the use of products and systems for protection and repair of concrete structures are based on a hierarchy of different levels, namely, options, principles, and methods. According to EN 1504-9, the following options shall be taken into account in deciding the appropriate action to meet the future requirements for the remaining service life of the structures:

1. Do nothing for a certain time but monitor;
2. Re-analyse the structural capacity, possibly leading to downgrading in function;
3. Prevent or reduce further deterioration;

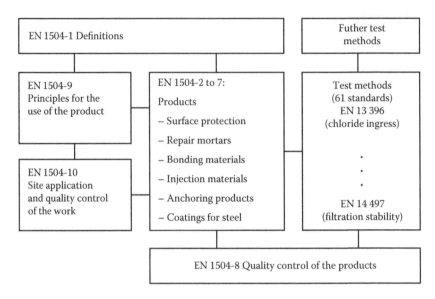

Figure 1.1 Structure of the series of European Standards EN 1504-1- to -10 for protection and repair of concrete structures.

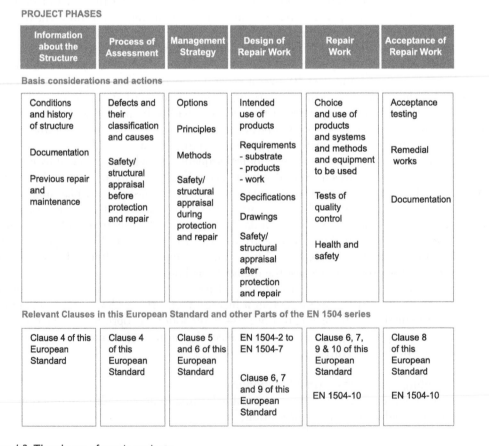

Figure 1.2 The phases of repair projects.

4. Strengthen or repair and protect all or part of the concrete structure;
5. Reconstruct or replace all or part of the concrete structure;
6. Demolish all or part of the concrete structure.

For protection and repair, different principles have been defined, separately for repair and protection of damages to the concrete and damages induced by reinforcement corrosion. Tables 1.1 and 1.2 show the six principles for protection and repair of concrete and the five principles to prevent damages due to reinforcement corrosion, respectively. These principles are based on RILEM Technical Recommendation 124-SRC, *Strategies for Repair of Concrete Structures Damaged by Steel Corrosion*. To protect or repair a concrete structure according to the principles, different methods are available. Altogether 43 methods are described within EN 1504-9. Not all of them are covered by the EN 1504 series, but by other standards, and some of them are actually not standardised.

The system of options, principles, and methods is the basis for the selection of products by the designer. The process of planning and selection of products is described within the next section.

Figure 1.3 shows the systematics of planning according to EN 1504-9. As already shown in Figure 1.2, the planning starts with the assessment of the status of the structure. Selection of options (repair strategy), repair principles, and repair methods are the following steps. The repair materials can be chosen based on this selection scheme.

EN 1504-9 defines performance characteristics for every repair method. The designer selects the performance characteristics based on the requirements of the special repair project and the selected repair methods. EN 1504-2 to –7 contain the performance characteristics of the products together with the corresponding test methods. This way the products are selected individually for the demands of the special case of repair or protection of a concrete structure. As a result, the products are described by a list of required performance characteristics instead of simple classes, resulting in a high level of flexibility for the designer.

Finally, inspection and maintenance requirements shall be defined by the designer.

Table 1.1 Principles for repair and protection for damages of the concrete

Principle no.	Principle and its definition
Prinicple 1 [P1]	Protection against ingress
Prinicple 2 [MC]	Moisture control
Prinicple 3 [CR]	Concrete restoration
Prinicple 4 [SS]	Structural strengthening
Prinicple 5 [PR]	Physical resistance
Prinicple 6 [RC]	Resistance to chemicals

Table 1.2 Principles for protection against reinforcement corrosion

Principle no.	Principle and its definition
Prinicple 7 [RP]	Preserving or restoring passivity
Prinicple 8 [IR]	Increasing resistivity
Prinicple 9 [CC]	Cathodic control
Prinicple 10 [CP]	Cathodic protection
Prinicple 11 [CA]	Control of anodic areas

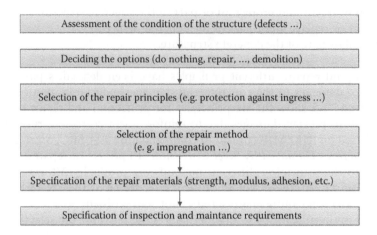

Figure 1.3 Systematics of planning according to EN 1504-9.

This systematic approach has also been used to structure this book. First, the different types of damages of the concrete and the reinforcement are described in Chapter 2. The theoretical background of the deterioration mechanisms is discussed for typical cases. It is essential to understand the process leading to the damages to be able to protect or repair concrete structures adequately.

Chapter 3 deals with the assessment of concrete structures, which is not regulated in the series of standards EN 1504, but is essential for successful protection and repair of concrete structures. Therefore, the methods are presented briefly that are usually used to test the condition of a structure directly on site and later using probes in laboratory. Chapter 4 explains briefly how the current and future conditions of the structure can be evaluated.

Chapter 5 describes the repair options according to EN 1504. In this chapter the materials that are usually used, their characteristics, and typical fields of application are described. The requirements of the repair products and systems according to EN 1504 Parts 2–7 regarding CE marking are not described in detail because they are listed in the standards.

Chapter 6 shows all principles and methods for protection and repair of concrete structures as given in EN 1504. Sketches and diagrams have been prepared by the authors to allow an easier understanding of all methods. For each method a short summary is given in the form of a table to get a quick overview of the method's special characteristics. This chapter is primarily important for the designers of repair works, but also for the people involved with execution, quality control, and the products for repair and protection of concrete structures.

Chapter 7 deals with execution of repair works and quality control, taking EN 1504-10 into account. For typical types of repair works, important steps and points need to be taken into account.

Maintenance of concrete structures is described in a chapter by itself (Chapter 8), because it is very important and essential for technically and economically optimal management of a structure. Every measure for protection and repair changes the condition of a structure that has to be implemented into the maintenance and management plans. Besides regular inspections, the use of monitoring systems may be helpful to give important information on the required maintenance.

Finally, in Chapters 9 and 10 an outlook and relevant literature/standards and guidelines are given.

Chapter 2

Deterioration mechanisms

2.1 CAUSES FOR DAMAGES ACCORDING TO EN 1504

Figure 2.1 shows the common causes of defects according to Figure 1 of EN 1504-9. With respect to later planning of repair, generally it should be distinguished between the defects in concrete and the defects caused by reinforcement corrosion. The purpose of the main assessment is:

- To identify the cause or causes of defects
- To establish the extent of defects
- To establish where the defects can be expected to spread to parts of the structure that are at present unaffected
- To assess the effect of defects on structural safety
- To identify all locations where protection or repair may be needed

More details on requirements for assessment are given in EN 1504-9.

2.2 DAMAGES TO THE CONCRETE

2.2.1 General

As already shown, EN 1504 generally separates damages of the concrete and damages induced by reinforcement corrosion. Therefore, in the following chapters, at first the damages of the concrete are described. Afterwards, the damages of the reinforcement are described, which can lead to damages of the concrete like cracks or spalls in a later stage of corrosion.

According to Table 3 in EN 1504-9, damages of the concrete may be induced by mechanical, chemical, or physical attack or by fire.

2.2.2 Mechanical attack

Mechanical attack may be induced by abrasion, fatigue of the material, impact overload, movement like settlement, explosion, or vibration. Especially for floors and roads, mechanical attack by pedestrians or vehicles has to be taken into account. Also, containers, structures for waterways, or rollways are exposed to specific mechanical attacks.

Figure 2.2 shows an example of the contact stresses induced by different types of tyres. It shows that steel tyres, e.g., used for transportation of goods, induce by a factor of about 100 higher stresses to the concrete surface than tyres filled with air.

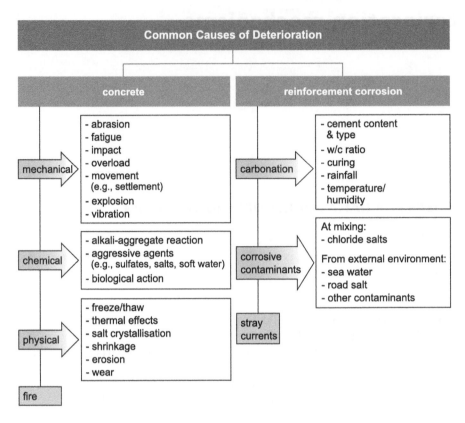

Figure 2.1 Common causes of defects according to EN1504-9.

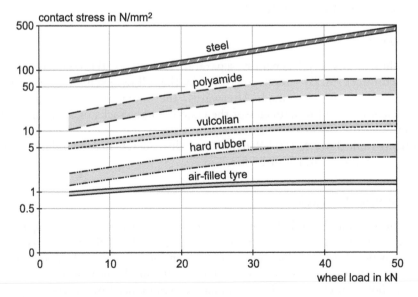

Figure 2.2 Contact stresses to concrete floors induced by wheels with different types of tyres. (From Sasse, H.R., Schäfer, H.G., Bäätjer, G., Borg, G., et al., *Schriftenreihe des Deutschen Ausschusses für Stahlbeton* (1996), no. 467.)

Figure 2.3 Example for a locally damaged concrete surface.

Figure 2.4 Example for damage induced by combined mechanical and corrosive attack.

Especially when the quality of a concrete surface is low, mechanical loads may lead to damages. Figure 2.3 shows local damage of the concrete surface of a car park deck. To avoid further growing of the damaged spot and accidents by people walking in the area, it should be repaired using a suitable mortar.

Figure 2.4 shows a damaged area in a parking garage that may be induced by different causes. It is obvious that reinforcement corrosion is active, but also frost or freeze-thaw attack might contribute to the damage. Furthermore, it seems that a local patch repair has already been carried out with inadequate quality.

The resistance of the concrete against mechanical attack is significantly controlled by the size and type of aggregates as well as the quality of execution (placement, compaction, and curing). If high mechanical loads are expected, the mechanical resistance of the concrete surface can be improved by screeds containing a special mix of hard aggregates.

2.2.3 Chemical attack

2.2.3.1 General

Chemical attack can be induced by different chemicals that are aggressive to concrete and result in dissolving and expanding attacks as described below. In most cases the binder of the concrete (cement stone) is attacked, but in certain cases the aggregates are also attacked.

2.2.3.2 Dissolving attack

A typical dissolving attack is induced by acids, which dissolve components of the concrete-forming soluble reaction products. As a result, the concrete loses strength, beginning from the surface where the acid is acting. In the course of time the strength is fully lost and the material disappears, reducing the concrete cover and the thickness of the concrete element. The corrosion rate depends on the type and concentration of the acid as well as the quality of the concrete. The dissolution rates can vary from some millimetres or even centimetres down to some microns per year. One sign of dissolving attack is when the concrete surface can be rubbed away like sand.

Acids may be present in concrete structures of the chemical industry, in sewage water, or in special cases in natural water, especially as carbonic acid or sulphurous acid.

Protection and repair of damages induced by dissolving acids requires special knowledge regarding acid-resistant repair systems, including solutions for joints, etc.

2.2.3.3 Expanding attack

In the case of an expanding attack, reactions are occurring within the concrete that form reaction products with a high volume. They are typically not soluble. An example for this corrosion mechanism is a sulphate attack at foundations from soils or groundwater. Besides sulphates, lime (CaO), magnesium (MgO), or magnesia salts can also induce expanding attacks.

Another example is the alkali-silica reaction (ASR), which occurs when certain types of aggregates or sand have been used in concrete that are not stable in alkaline environments at high pH values. Then the amorphous silicic acid particles of the aggregates react with the alkalis of the cement stone. This reaction leads to the formation of products with a high volume, resulting in an inner pressure within the concrete. The expansion can develop to such an extent that the inner structure of the concrete cracks and loses strength.

Figure 2.5 shows a photograph of a cut section of concrete suffering from ASR. The concrete sample has been impregnated with a green resin to make cracks and voids visible. It can clearly be seen that a crack runs directly through an aggregate grain. This highlights the high inner pressure occurring during ASR.

If a critical level of inner pressure is reached, macrocracks occur and the problem of ASR gets visible. Figures 2.6 and 2.7 show examples of ASR. Often a gel can be found within the cracks induced by ASR.

Local pop-outs can occur, when aggregates contain critical amounts of sulphides, e.g., pyrite (FeS_2) or pyrrhotite (FeS), which can oxidise. Figure 2.8 shows a pop-out induced by an aggregate containing pyrite. As long as the number of pop-outs is quite low, as a repair method the critical aggregates can be drilled out and replaced by a suitable mortar. To reduce the risk of further pop-outs, a surface protection system may be applied that reduces the water content of the concrete down to an uncritical level (see Method 2.3 in Section 6.3.4).

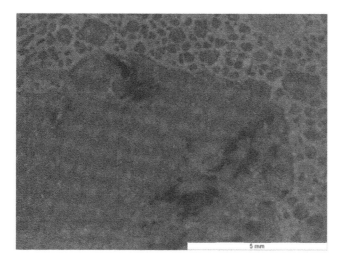

Figure 2.5 ASR-induced crack within an aggregate grain.

Figure 2.6 Example of crack formation induced by ASR with typical formation of a gel within the crack (pillar in Scandinavia).

Pop-outs can also be induced by single aggregates, which are not sufficiently resistant against frost or freeze-thaw attack. However, pop-outs induced by pyrite can easily be realised by the brown colour looking like rust. To get a clear diagnosis, a chemical analysis of the corrosion products and the aggregates can be carried out.

2.2.4 Biological acid attack

Biological acid attack often occurs in sewage treatment plants or sewage ducts. One of the most aggressive types, which often occurs under anaerobic conditions, is the biogenous sulphuric acid attack. For this type of corrosion, microorganisms play an important part because they produce the sulphuric acid. The corrosion rates often are extremely high and in the range of 1 cm per year. Figures 2.9 and 2.10 show an example of a sewage duct in Germany after a service life of only six years. It can be seen that the aggregates can be moved

Figure 2.7 Example of crack formation induced by ASR in a progressed state (column in South Africa).

Figure 2.8 Pop-out at a concrete surface induced by aggregates containing pyrite.

out of the concrete surface by hand. Figure 2.10 depicts a core drilled out of the structure, showing that significant amounts of the cement stone are already dissolved and disappeared.

This biogenous sulphuric acid attack typically occurs in the upper areas of the duct above the sewage water level, where microorganisms produce a highly concentrated acid.

For the repair of such damages, it has to be considered that the acid attack has already penetrated deeper into the concrete. Therefore, usually the upper concrete layer has to be removed until a depth is reached where the strength is not decreased by the ingress of the acid. The critical depth can be investigated by microscopic and chemical analysis of cores taken from the structure (see Section 3.2.6.2).

The selection of a suitable repair material has to be carried out carefully because it must be acid resistant, when the environmental conditions of the structure cannot be improved, and a sufficient and durable bond to the substrate is required.

Figure 2.9 Concrete surface of a six-year-old sewage duct damaged by biogenous sulphuric acid attack.

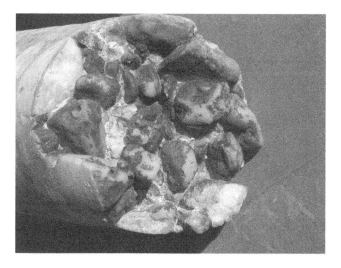

Figure 2.10 Core drilled from the structure shown in Figure 2.9.

2.2.5 Physical attack

2.2.5.1 General

According to EN 1504-9 physical attack can be induced by freeze-thaw, thermal effects, salt crystallisation, shrinkage, erosion, or wear. These attacks often lead to cracking or spalling. In extreme cases, frost attack may destroy young concrete totally, when it is placed in winter without thermal protection.

2.2.5.2 Frost and freeze–thaw attack

Frost and freeze-thaw attack can be prevented by using the well-known rules of concrete technology. However, the frost resistance of older structures sometimes is not high enough due to several reasons (bad mix design, bad quality of execution, changing of the environmental conditions, etc.).

If the frost damage has progressed quite far, usually the damaged surface layer has to be removed and replaced by a suitable mortar or concrete. In an early stage, the application of a surface protection system reducing the water content of the concrete may be sufficient, provided that a good bond to the substrate can be ensured.

2.3 DAMAGES TO THE REINFORCEMENT

2.3.1 General

Reinforcement corrosion can be induced by carbonation, corrosive contaminants like chlorides, or stray currents. To be able to select the optimal repair method and materials it is important to understand the mechanisms of corrosion. Therefore, in the following sections the theoretical background of corrosion of steel in concrete is explained.

Generally steel is protected in sound concrete by the alkalinity of the concrete with pH values typically higher than 13. This effect is demonstrated in Figure 2.11. Two reinforcing bars have been put into different solutions: one in tap water and one into saturated lime solution in water $(Ca(OH)_2)$ with a pH value of 12.6. As shown in Figure 2.11, the rebar in tap water corrodes and the colour of the water gets rusty. However, the rebar in the lime solution shows no signs of corrosion.

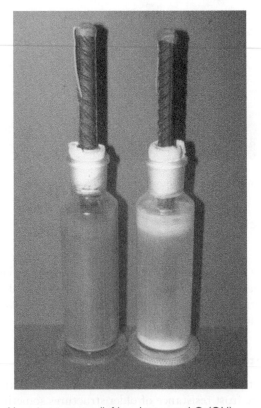

Figure 2.11 Reinforcing steel bars in tap water (left) and saturated $Ca(OH)_2$.

This effect can be explained by the formation of a passive layer on the steel surface, which prevents the further dissolution of iron. This passivity occurs at pH values above about 9–10 and leads to a corrosion behaviour compared to that known from a stainless steel at the atmosphere.

As long as the passive layer is active, no corrosion occurs. However, there are at least two main processes that can destroy the passive layer (depassivation):

- Carbonation of the concrete, i.e., the reaction of the alkalis in the pore solution with CO_2 within the air, leads to a decrease of the initial pH value of >13 to values of <9.
- Chloride ingress into the concrete can cause depassivation of the steel surface, where a critical chloride content is reached.

Based on this, according to Tuutti (1982) the corrosion of steel in concrete can be devided into two phases: the initiation and propagation phases (see Figure 2.12). During the initiation phase carbonation or chloride ingress occurs. As long as the carbonation depth or the depth of the critical chloride content has not reached the level of the reinforcement, the steel reinforcement is passive and no deterioration occurs.

The deterioration phase starts when depassivation occurs by carbonation or chlorides. Corrosion proceeds until a certain limit for the deterioration has been reached and repair measures are required.

Over the course of time, when there are visible signs of corrosion cracking and spalling occurs, it is due to the fact that rust has a higher volume than steel. The volume factor

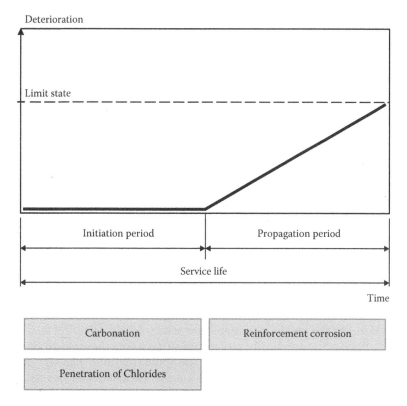

Figure 2.12 Schematic representation of the corrosion process of steel in concrete. (From Tuutti, K., *CBI Research* (1982), no. fo. 4:82.)

(volume of the rust/volume of the steel) depends on the environmental conditions and varies in a wide range. Typical corrosion products and volume factors are:

- Fe_3O_4 (black): 2.1
- Fe_2O_3 (red brown): 2.14
- $Fe_2O_3*H_2O$ (yellow): 3.12
- $Fe(OH)_2$ (white): 3.71
- $Fe(OH)_3$ (red brown): 4.82
- $Fe_2O_3*3H_2O$ (red brown): 6.5

When the concrete is very wet and its water content is close to saturation, e.g., in wet soil, often black rust is found, which has a low volume factor and does not induce high pressure in the concrete. If the water content of the concrete is quite low and considerable amounts of oxygen are available, oxide-rich rust will be formed, which induces high pressure in the concrete, leading to quick spalling. Typically, the volume factor is in the range between two and four for outdoor conditions with regular rainfalls and dry periods.

For chloride-induced corrosion the rust products may be formed apart from the corroding area (see Section 2.3.4), especially in wet concrete. In such cases high steel removal rates can occur without spalling. From the viewpoint of safety this is unfavourable, because there are no visible signs of the corrosion process even in a progressed state.

To be able to plan adequate repair measures, it is necessary to understand the corrosion mechanisms. This is especially important in the case of chloride corrosion, stray current-induced corrosion, or for the use of electrochemical methods for protection and repair.

2.3.2 Electrochemical background

Corrosion of steel in concrete is an electrochemical process. Due to extensive research in the field of corrosion during the last decades, today the mechanisms are quite clearly understood. The electrochemical nature of steel corrosion in concrete is used for the diagnosis of structures (potential mapping), embedded corrosion sensors (electrical current measurements), and the calculation of corrosion rates.

Looking at a corroded steel surface, only rust is visible. After removal of the rust it becomes visible that a part of the steel is missing and the cross section of the steel has been reduced. Corrosion research has shown that these macroscopic effects are the result of processes occurring on the molecular level. Figure 2.13 explains schematically how corrosion induced by carbonation of the concrete proceeds.

One requirement for electrochemical corrosion is the presence of an electrolyte at the steel surface. For steel in usual outdoor conditions this is given by a thin film of water containing different types of ions. In the case of steel in concrete the electrolyte is given by the aqueous solution within the pore system of the concrete. As already explained, the high pH value of this pore solution causes passivation of the steel.

However, when depassivation occurs, e.g., by carbonation of the concrete, the electrochemical process of corrosion starts. This consists of an anodic oxidation and a cathodic reduction reaction. At the anode iron ions enter into solution releasing two free electrons within the steel. Besides the anodically acting steel surface area these electrons react together with water and oxygen to hydroxide ions OH^- (cathodic reaction). The OH^- ions then react together with the Fe^{++} ions to rust. Depending on the availability of oxygen, more or less oxygen is present in the rust products, as explained in the previous Section (2.3.1). In all

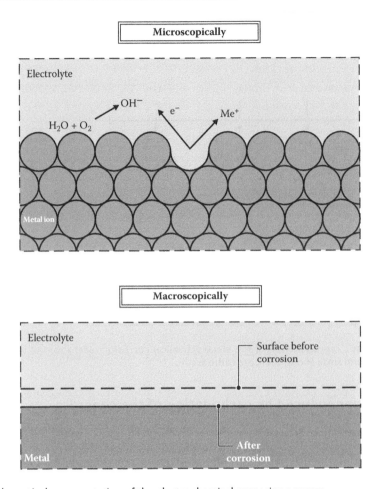

Figure 2.13 Schematical representation of the electrochemical corrosion process.

areas where the steel surface is depassivated and sufficient oxygen and water are available this process takes place, leading macroscopically to a reduction of the cross section of the steel, as shown in Figure 2.13.

Often only a part of the steel surface is depassivated. In such cases the anodic reaction can only take place at the depassivated areas. The remaining areas react cathodically. This is typical for chloride-induced corrosion. The area where the critical chloride content is reached at the steel surface at first will depassivate and start to act as an anode. The areas beside this anode will act as cathodes. However, cathodically acting areas are to a certain level protected against corrosion by cathodic protection and will not depassivate as quickly as the first anode.

This corrosion process with local anode and large cathodes is schematically presented in Figure 2.14. When cathodic areas are large compared to the anodic areas, the corrosion reaction at the anode is accelerated according to the well-known area rule of corrosion (see also Section 2.3.4). The mechanisms of chloride-induced corrosion are described more in detail after carbonation-induced corrosion.

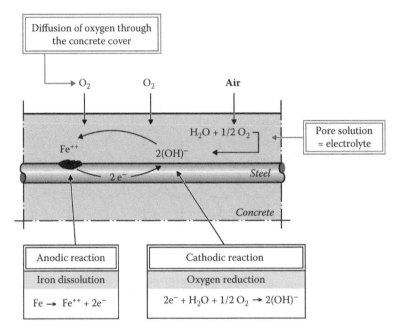

Figure 2.14 Schematic representation of the electrochemical corrosion mechanism of localised depassiv-ation with small anode and large cathodes.

2.3.3 Corrosion induced by carbonation of the concrete

Carbonation-induced corrosion of steel in concrete is probably the most prevalent type of reinforcement corrosion. It occurs at every concrete surface when the carbonation front has reached the level of the reinforcement. During recent years, new concrete structures are designed in a way that the carbonation front will not reach the steel surface within the design service life, e.g., 50 years. Old buildings often show this type of corrosion due to insufficient thickness or quality of the concrete cover. The extent of corrosion depends considerably on the water content of the concrete. This is shown later in several examples.

The first visible signs of carbonation-induced corrosion are typically cracks forming in the concrete over the corroding reinforcement. Figure 2.15 shows an example where the position of the vertical reinforcement bars can be seen indirectly by the cracks and spalls. A comparison with a cover meter shows that these cracks run alongside every steel bar with a small concrete cover.

The second stage is shown in Figure 2.16. The concrete over the corroding steel has fallen down. Obviously the steel stirrups have been placed into the concrete without spacers at the concrete surface. After the concrete cover has disappeared, the corrosion rate of the rein-forcement is in the same range as that of steel exposed to the atmosphere.

Depending on the environmental conditions and the concrete quality, the spalling of the concrete above the corroding reinforcement can also occur in a progressed corrosion state. Figure 2.17 shows an example where the corrosion has proceeded quite far at a parapet exposed to outdoor conditions in Germany. Figures 2.18 and 2.19 show examples of cor-roded reinforcement in different levels of corrosion.

Carbonation-induced corrosion also often occurs when the concrete quality locally is low even when the thickness of the concrete cover is according to the standards, e.g., in cases

Figure 2.15 Cracking and spalling due to carbonation-induced corrosion of the vertical reinforcement within a concrete façade.

Figure 2.16 Freely exposed reinforcement due to carbonation-induced corrosion.

of bad compaction or gravel pockets. Figure 2.20 shows an example where the distance between the reinforcing bars has been too small so that the concrete could not pass and embed the reinforcement from all sides. In such cases the carbonation front can quickly reach the steel surface and the corrosion rate is mainly determined by the humidity of the environment and the water content of the concrete.

In outdoor structures the rate of carbonation-induced corrosion may be low when the concrete is sheltered, e.g., by a roof. Figure 2.21 shows an example of a bridge where high corrosion rates only occur in areas where water runs over the concrete surface by a defect in the drainage system. The remaining reinforcement is not corroding heavily because it is sheltered from rain by the bridge deck.

Figure 2.17 Carbonation-induced corrosion at a parapet.

Figure 2.18 Rebar with a reduced cross section by carbonation-induced corrosion.

Usually under indoor conditions the concrete is quite dry and the carbonation rate is high. However, due to the limited amount of water in the concrete the corrosion rates are very limited. Figure 2.22 shows the reinforcement of a column within a building in the area of a gravel pocket that is roughly 50 years old. It can be assumed that the carbonation front has reached the level of the reinforcement very quickly, but the extent of corrosion is not significant. However, the gravel pocket should be repaired to improve the bond between reinforcement and concrete and the fire resistance.

Another example is shown in Figure 2.23. The reinforcement of a roughly 100-year-old storage building shows brown rust on the surface, but no considerable loss of cross section. This effect can only be explained by a low humidity throughout the whole service life. Of course this structure needs concrete repair.

At the same structure the reinforcement is heavily corroded in an area where water could penetrate through the roof due to defects in the sealing (see Figure 2.24). Obviously

Figure 2.19 Carbonation-induced corrosion in a progressed state.

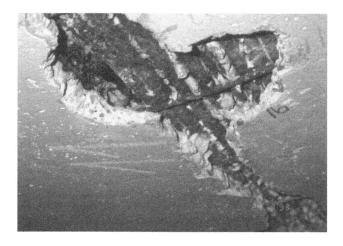

Figure 2.20 Reinforcement corrosion in the area of a large gravel pocket. The remaining concrete cover has been removed.

this leakage has been active for a minimum of some years, resulting in such high losses of cross section. In such cases the load-bearing capacity of the structural element needs to be checked, and usually strengthening works are required.

Corrosion damages are also often found where repair or protection works have not been carried out in a technically correct manner. Figure 2.25 shows an example where the concrete cover has spalled due to corrosion of the reinforcing bar even though the concrete has been coated. This could be caused by bad preparation of the concrete surface before coating.

The corrosion of the reinforcement shown in Figure 2.25 could have been induced by carbonation, but also possibly by chlorides from de-icing salts. Chloride-induced corrosion of steel in concrete is based on different mechanisms, which are explained in Chapter 3.

Figure 2.21 Local damages due to corrosion in the area of a defect in the drainage system.

Figure 2.22 Reinforcement of a roughly 50-year-old column in the area of a gravel pocket with no significant signs of corrosion.

2.3.4 Chloride-induced corrosion

Chloride-induced corrosion of steel in concrete does not usually lead to a uniform loss in the cross section of the reinforcement like carbonation-induced corrosion, but to localised corrosion spots. This is due to the fact that a more or less highly concentrated hydrochloric acid (HCl) emerges at the anodes, which causes high corrosion rates and stabilises the corrosion process.

Figure 2.26 shows schematically the mechanism of the so-called pitting corrosion. Negatively charged ions like chlorides migrate to the anode and produce HCl. The passive steel surface areas besides the anodes are acting as cathodes, and to a certain extent they are cathodically protected. This leads to the effect that the anode grows more into the direction to the inner part of the steel than to the sides along the surface. Due to this form of iron removal the type of corrosion is called pitting. In the course of time several small pits may

Figure 2.23 Reinforcement of the girders of a roughly 100-year-old storage building with no considerable strength loss.

Figure 2.24 Damage of a concrete roof by carbonation-induced corrosion in the area of a defect sealing.

grow together, forming larger hollows. Over the pits layers of rust are formed that separate the pit with low pH values from the concrete with high pH values.

Figure 2.27 shows as an example the surface of a reinforcing bar probe, which has been embedded over two years in concrete with 2 wt% of chlorides. Upon removal from the concrete the steel was cleaned by an acid solution containing corrosion inhibitors to remove the rust products. After cleaning the steel loss becomes visible. It can be seen that a hole with a diameter of about 3 mm and a depth of about 3 mm has developed.

In practice such pits are covered with rust and are only visible after removal of the rust. Figure 2.28 represents schematically the difference between uniform corrosion induced by carbonation and chloride-induced macrocell corrosion with local pits.

Usually macrocells with large cathodes and small anodes occur, resulting in high corrosion rates. Figure 2.29 shows a stirrup taken from a concrete structure in the Middle East after a service life of only 4 years. At the bottom parts (Figure 2.29) a crack crossed the stirrup, where saltwater could penetrate quickly to the steel surface. As the initial diameter of the stirrup was 8 mm, the corrosion rate must have been higher than 1 mm every year.

Figure 2.25 Spalling of the concrete cover at a coated concrete wall.

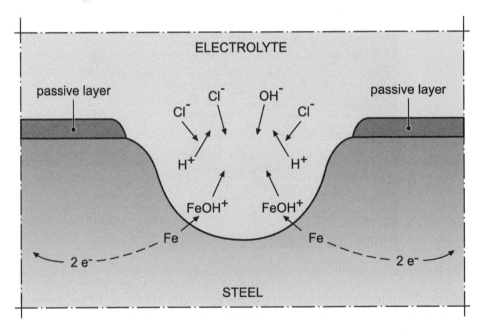

Figure 2.26 Schematic representation of the conditions within an anodic pit induced by the action of chlorides. (From Page, C.L., Havdahl, J., *Materiaux et Constructions* 18 (1985), no. 103, s. 41–47.)

Such high corrosion rates can only be explained by the formation of macrocells with localised anodes and large cathodes.

Chloride-induced macrocells in practice can occur in different forms. Figure 2.30 shows an example of corrosion in chloride-contaminated soil, where the existence of a macrocell has been demonstrated by targeted investigations. The reinforcement of a retaining wall in Berlin was corroded heavily in the area of badly compacted concrete exposed to chloride-contaminated soil. However, the corrosion products have not been found close to the anodes, but significantly above the corroding areas in the ground. This leads to the conclusion that

Figure 2.27 Reinforcing steel with a chloride-induced pit after pickling.

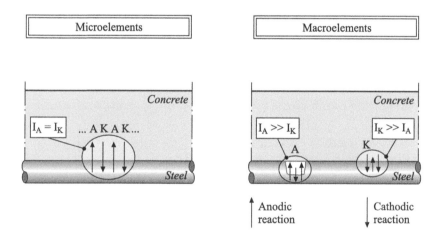

Figure 2.28 Uniform corrosion with local corrosion cells (e.g., due to carbonation) and macrocells with localised anodes and cathodes (e.g., due to chlorides).

a macrocell has been active with large cathodes at the well-aerated zone above the ground. In the electrical field between the anode and the cathode positive ions migrate toward the cathode, explaining the presence of ruse in the soil above the corroding areas.

One consequence of the formation of macrocells is that high corrosion rates can occur where they are not expected. Figure 2.31 shows the situation of a defect sealing on a bridge deck. It could be assumed that the possible corrosion rates are small because the availability of oxygen near the anode is small. However, the reinforcement of the opposite side of the bridge deck is well aerated and will act as a large cathode accelerating corrosion at the upper reinforcement depassivated by chlorides.

The electrical connection between all reinforcing bars is usually given within one concrete structure by the connections between the different layers of the steel reinforcement.

From these examples it can be concluded that the diagnosis of damages caused by choride-induced corrosion requires careful investigations. However, to plan effective repair measures,

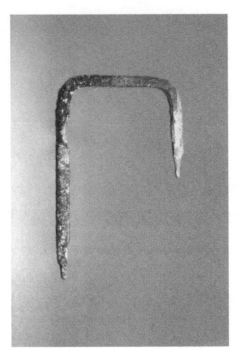

Figure 2.29 Four-year-old stirrup taken from cracked concrete exposed to seawater.

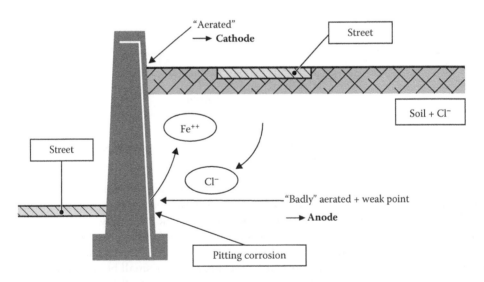

Figure 2.30 Schematic drawing of a retaining wall where the presence of a macrocell has been demonstrated (Laase and Stichel 1983).

it is important to evaluate the damage mechanisms as exactly as possible. In Figures 2.32 to 2.34 typical damages caused by chloride-induced corrosion are shown. Specifically, examples of damages in parking garages induced by chloride-contaminated water running through defect joints are shown. When the inspection intervals are quite long, such damages may be undetected until they are in a progressed state. In such cases extensive measures are

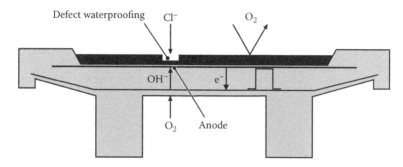

Figure 2.31 Sketch of a possible macrocell caused by chloride-induced corrosion on a bridge deck.

Figure 2.32 Example for corrosion damages with spalling of the concrete in a parking garage induced by leaking joints.

required, including diagnosis, checking of the load-bearing capacity, and planning of repair measures. In certain cases it can be necessary to close the structure until a safe use can be stated. Often strengthening measures are required.

Besides aspects of load-bearing capacity and durability, the usability and safety also have to be taken into account. Figure 2.35 shows an example of a piece of concrete that has fallen down from the corroding area shown in Figure 2.34. When signs of spalling like cracks, rust spots, or deformations are visible, the concrete surface should be checked using the simple hammer method or more advanced methods (see Chapter 3). The weak parts of concrete have to be removed immediately to ensure safety. Furthermore, in such cases the consequences of the defects on the structural safety need to be checked.

A typical damage in parking garages is the corrosion of the feet of columns and walls by sucked-up water containing de-icing salts. Figure 2.36 shows an example where chlorides from de-icing salts have penetrated into the foot of a column and caused corrosion-induced spalling. Typically, roughly about 20 to 50 cm is affected by this problem.

Figure 2.33 Example for corrosion damages in a parking garage induced by leaking joints in critical areas regarding load-bearing capacity.

Figure 2.34 Example for corrosion damages in a parking garage induced by missing sealants of the building joints in critical areas regarding load-bearing capacity.

For repair of such damages the load-bearing capacity of the column has to be considered. Often it is not possible to remove the entire damaged concrete because the cross section of the column cannot be reduced too much, as it is required to bear the vertical loads from the structure. Therefore, the static reserves have to be proved in the phase of repair design before execution of any works on site. Usually a limited loss of cross section is possible. However, it has to be controlled carefully during removal of the concrete, that the conditions are kept. However, in some cases, a temporary additional column has to be installed to ensure safety during the repair works.

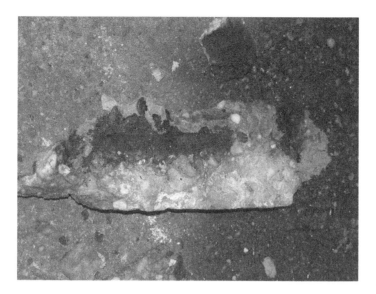

Figure 2.35 Piece of concrete fallen down from the corroding area shown in Figure 2.34.

Figure 2.36 Spalling of the concrete caused by reinforcement corrosion induced from de-icing salts.

When concrete is directly exposed to seawater or close to the sea it is also attacked by chlorides from the seawater or winds carrying chlorides from the seawater. To prevent damages high quality and thickness of the concrete cover are essential. Often additional measures like coatings are used for protection. For older structures, often quality and thickness of the concrete cover are not sufficient, and corrosion problems can be found. Usually the corrosion rates are very high because compared to the action of de-icing salts, seawater is attacking the structure continuously.

Figures 2.37 and 2.38 show examples for damages of concrete structures exposed to seawater. In the example shown in Figure 2.38 the access of the structure is forbidden due to the risk of a collapse.

Figure 2.37 Concrete structure damaged by seawater-induced reinforcement corrosion.

Figure 2.38 Concrete pier closed to the public due to seawater-induced reinforcement corrosion.

2.3.5 Corrosion mechanisms in cracks

In cracks aggressive substances like chlorides can directly penetrate into the concrete very quickly. Experience has shown that reinforcement corrosion starts quickly and with high rates, especially when wide cracks reach down to the level of the reinforcement or cross the reinforcement. The most unfavourable situations are when a crack runs through the whole cross section of a concrete structure.

Due to local depassivation of the reinforcement in the area of the crack, often large macrocells occur with small anodes in the crack area and large cathodes in the areas between the cracks. This situation is shown in Figure 2.39. As already described previously, such macrocells lead to high corrosion rates (see also Figure 2.29).

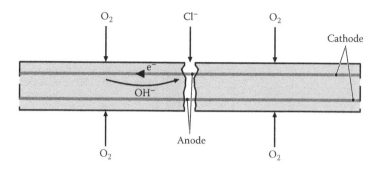

Figure 2.39 Schematic representation of the mechanism of macrocell corrosion in the area of a crack in concrete.

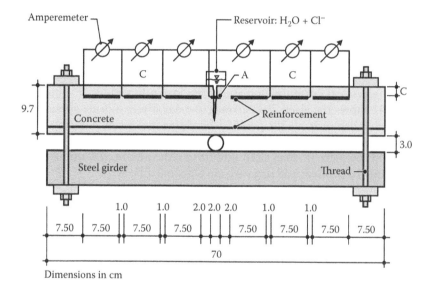

Figure 2.40 Laboratory test specimen to investigate the corrosion mechanism in the area of a crack in concrete. (From Schießl, P., Raupach, M., *Bauingenieur* 69 (1994), no. 11, pp. 439–445.)

To verify this mechanism laboratory tests have been carried out using the electrochemical nature of corrosion (Schießl and Raupach 1994). To simulate a long reinforcing steel bar, a segmented steel bar consisting of several pieces of reinforcing steel has been used, as shown in Figure 2.40. All pieces of steel have been connected to ampere meters to allow a measurement of the electrical currents flowing along the segmented bar. According to Faraday's law, the electrical current flow behaves proportionately to the mass loss by corrosion. That means the level of electrical currents indicates the corrosion rate.

To generate a crack in concrete, the concrete beam has been clamped against a steel girder. By threads at the ends of the beam and a steel cylinder in the centre, bending forces have been applied to the concrete, resulting in high tensile stresses on top on the beam in the centre area. The loads have been increased until the concrete cracked with a defined crack width of 0.5 mm.

The electrical currents between the steel bars have been continuously measured. After generating a crack a chloride solution has been applied to the crack to induce corrosion.

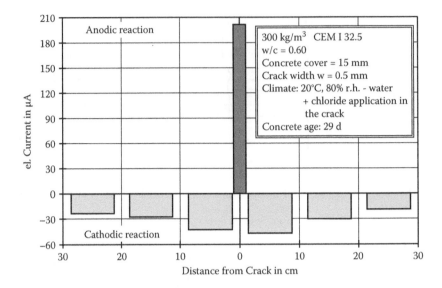

Figure 2.41 Electrical currents measured at the specimen shown in Figure 2.40.

Immediately the electrical currents increased from insignificantly low values to quite high levels. Figure 2.41 shows the electrical currents one day after chloride application.

The results clearly demonstrate the presence of a large macrocell. The sign of the electrical currents shows the direction of the current flow. A positive sign means that more electrons are moving out of the piece of steel than into it; i.e., the anodic reaction is dominant, and vice versa. The sum of positive and negative currents must be zero.

Figure 2.41 shows that a small anode occurs in the area of the crack as expected, and the rest of the reinforcement reacts cathodically. As to be expected, the cathodic reaction rates decrease with the distance from the anode. In a distance of about 30 cm from the crack, a still rather high cathodic action has been measured.

This result means that the reinforcement between the cracks acts as large cathode accelerating the corrosion rate of the reinforcement in the area of the crack in concrete. It confirms the macrocell corrosion mechanism shown in Figure 2.39.

For the conditions of concrete structures in practice, it can be concluded that cracks are very unfavourable regarding chloride-induced reinforcement corrosion: first, they allow a quick chloride ingress and initiation of corrosion, and second, they lead to macrocell corrosion with high corrosion rates.

Figures 2.42 and 2.43 show cracks running through a deck of a parking structure at the bottom side. They are also visible on the top side of the deck. Water contaminated with chlorides can run through the cracks, causing corrosion of the reinforcement.

To evaluate the condition of the reinforcement in the area of the crack a piece of steel has been removed and cleaned from rust by acid treatment. The crack width in concrete is 0.8 mm. This is significantly higher than usual. In most cases the calculated crack width limitation of the concrete by the reinforcement is about 0.3 mm. The age of the structure was roughly 10 years.

The position of the crack in concrete can clearly be identified in the right half of Figure 2.44. The remaining cross section is only about 25%. On the safe side, the load-bearing capacity of the steel bar in the area of the crack will be estimated to zero.

Figure 2.42 Reinforcement corrosion in the area of a crack at the bottom of the deck of a parking structure.

Figure 2.43 Reinforcement corrosion in the area of a crack at the bottom of the deck of a parking structure.

Figure 2.44 Reinforcing steel bar taken from the area of a 0.8 mm wide crack in a roughly 10-year-old parking garage in Germany.

Figure 2.45 Crack closed by self-healing effects of the concrete.

This example demonstrates again the detrimental effect of cracks in concrete on the corrosion behaviour. However, to estimate the corrosion behaviour in the areas of cracks, an individual assessment of the special conditions of every structure is required taking all relevant parameters from the materials, construction, and environment into account.

Figure 2.45 shows an example where the crack has been closed by self-healing effects. If self-healing occurs shortly after initiation of the crack, no effect of the crack regarding reinforcement corrosion has to be expected. If no information on the history of the crack is available, carbonation depth and chloride content of the concrete within the cracks should be checked to enable evaluation of the corrosion risk.

2.3.6 Corrosion induced by leaching out of the concrete

Reinforcement corrosion can also be induced in the area of cracks or joints where water passes through over significant periods of time. This can lead to a local reduction of the pH value where the alkali ions are transported away with the water. Figure 2.46 shows a crack within a retaining wall as an example. Due to water flowing through the crack for many years, corrosion has been initiated in the area of the crack and proceeded so far that most of the reinforcing bars have been torn apart after about 40 a. The chloride contents in the concrete are so low that chlorides can be excluded as a reason for depassivation. Carbonation

Figure 2.46 Corrosion of the reinforcement in the area of a water-bearing crack induced by leaching out of the concrete. (Photo taken by Consulting Engineer, Raupach, Bruns.)

Figure 2.47 Detail of Figure 2.46 after spraying a phenolphthalein solution on the concrete surface showing reduced pH values (no red colour) in the area along the crack.

cannot be significant in the area of mostly wet concrete. The leaching-out effect can be proven by the phenolphthalein test, as shown in Figure 2.47. If the pH value is reduced so much that corrosion can start, the colour will not turn to pink, which would indicate high pH values as for young concrete, but no colour change will be visible.

2.3.7 Corrosion induced by stray currents

Besides carbonation or chloride ingress, stray currents may induce corrosion of the reinforcement. This type of corrosion only occurs in very exceptional cases, where electrical current

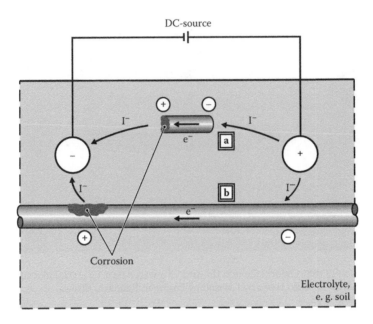

Figure 2.48 Mechanism of stray current corrosion of the reinforcement in concrete at a local structure (a) or a continuous structure (b).

fields are active in the ground and the usual measures (grounding) to prevent stray current corrosion are missing or defective.

Typical sources for electrical currents in the ground are electrical railroads or trams, electricity cables, and facilities for welding or electrolysis. However, to prevent problems with stray currents, electrical installations have to be installed in a way that the amount of possible stray currents is limited.

Nevertheless, if high stray currents are present in the ground, they can induce corrosion of the reinforcement of concrete structures within the electrical field. Figure 2.48 shows the mechanism of stray current corrosion at concrete at a local structure, e.g., a foundation (a), or a continuous structure, e.g., a pipeline or duct (b).

The stray current always attempts to take the short path through the well-conductive metal (e.g., steel reinforcement) instead of through the ground. The area where the current enters into the metal is cathodically protected, but the area where it runs out of the metal and back to the ground is anodically loaded and subject to corrosion.

If corrosion problems at the reinforcement in the ground are difficult to explain and when strong AC or DC sources are present, it should be checked whether stray currents may be the reason for reinforcement corrosion. However, this requires special instruments and knowledge and should only be executed by experts.

2.3.8 Corrosion mechanisms of prestressing steels

In prestressed concrete structures, special corrosion mechanisms may occur, like stress corrosion cracking or hydrogen embrittlement. These corrosion mechanisms are influenced not only by the environmental conditions, but also by the level of tensile stresses within the steel. Furthermore, the resistance of the prestressing steels against stress corrosion

Figure 2.49 Uncorroded prestressing steels in well-compacted grout after opening of the ducts by high-pressure water jetting.

cracking depends on the chemical composition and metallurgical condition, which again are influenced by the production technology. A detailed description of the mechanisms would go beyond the constraints of this book.

The quality of the grout or mortar around the prestressing steel plays an important role for the risk of stress corrosion cracking. To assess this risk, usually the prestressing steels are opened, e.g., using high-pressure water jetting (see Section 5.2). After opening of the ducts the degree of compaction of the mortar can be seen and evaluated. Usually the mortars are well compacted, but also areas may be found where the compaction is bad. Figures 2.49 and 2.50 show examples of areas where the ducts of the prestressing steels of two bridges in Germany have been opened, and in one case well (Figure 2.49) and in the other case bad (Figure 2.50) compacted mortar has been found. In the latter case rust is already visible. Due to the uncontrollable risk of sudden failures by stress corrosion cracking, the structure had been demolished.

For the evaluation of the conditions of prestressed concrete structures, generally the assessment of the load-bearing capacity is required. The assessment of the condition regarding the risk of stress corrosion cracking shall only be carried out by experienced experts taking the individual conditions of the structure into account, e.g., type of prestressing steel, structural details, level of static and dynamic loads, material for the grout or mortar, environmental condition, and cracks.

Figure 2.50 Corroded prestressing steels in badly compacted grout after opening of the ducts by high-pressure water jetting.

Chapter 3

How to assess the status of a structure

3.1 SIGNIFICANCE

Depending on the structural condition at the time a survey is conducted, it might seem that a limited assessment of the status is sufficient, but even when only local damages are visible, it can be necessary to investigate the entire structure. The decision to what extent the assessment has to be carried out will usually be made on the basis of a preliminary visual inspection, as well as the experience of the engineer. However, it can be necessary to adapt the scheduled investigations on site due to the results achieved during the survey. In all cases the assessment of the structure's condition has to be done with high accuracy because the results are the basis for selecting the specific repair options as well as repair methods. Figure 3.1 shows an example of a time-consuming assessment in which the load-bearing capacity of the structure was severely impaired by chloride attack, which is the need for this complicated assessment. The complexity of the assessment was due not only to the size of the structure, but also to the location of the structure. The structure is located in the tidal zone on the German coast, so all investigations had to be done with large rigs instead of boats, which limited the possible times of assessment.

The following sections cover selected common diagnosis methods, which are used to assess the status of a concrete structure.

3.2 DIAGNOSIS

3.2.1 General procedure

According to EN 1504-9:

> An assessment shall be made of the defects in the concrete structure, their causes, and of the ability of the concrete structure to perform its function.
>
> The process of assessment of the structure shall include but not be limited to the following (EN 1504):
>
> a. the visible condition of the existing concrete structure;
> b. testing to determine the condition of the structure and reinforcing steel;
> c. the original design approach;
> d. the environment, including exposure to contamination;
> e. the history of the concrete structure, including environmental exposure;
> f. the conditions of use, e.g., loading or other actions;
> g. requirements for future use.

Figure 3.1 Assessment of a concrete structure under complicated conditions.

It is important to mention that this list of steps is not comprehensive and might be expanded depending on the specific structure as well as type of damages.

As long as the assessment is carried out in the so-called initiation phase, usually no visible damages can be detected, and it can be decided whether protective measures are necessary to reach the designated lifetime of the structure. Also, if no damages due to reinforcement corrosion are visible, a survey can save costs because the required repair as well as the protective measures are not as labour-intensive as the required repairs after concrete damages have occurred (Polder et al. 2013). If an assessment is carried out during the corrosion phase, usually not only future measures that should ensure the designated life span, but also immediate measures that should secure the load-bearing capacity of the structure are subject of an assessment. Thus, it is important to know the status of the structure as well as the goal of the assessment in order to be able to conduct a complete survey of the structure.

Figure 3.2 gives an overview of procedures for the assessment of the condition of concrete structures regarding reinforcement corrosion (Raupach et al. 2013).

3.2.2 Concrete quality

3.2.2.1 General

The judgement of the concrete quality should be done under consideration of the function of the concrete within a structure. The border zone of the concrete is responsible for any ingress protection of a concrete structure against harmful substances, such as chlorides, carbon dioxide, or sulphur. The concrete's core is mainly responsible for the load-bearing capacity of the structural element.

3.2.2.2 Defects and voids

Defects and voids within a concrete structure should always be recorded while investigating the entire structure by visual inspection. Mainly old buildings show distinct areas in which the concrete is not compacted properly, as depicted in Figure 3.3. Such areas may be visible from the outside. But if these areas are hidden and it is obvious or expected that a

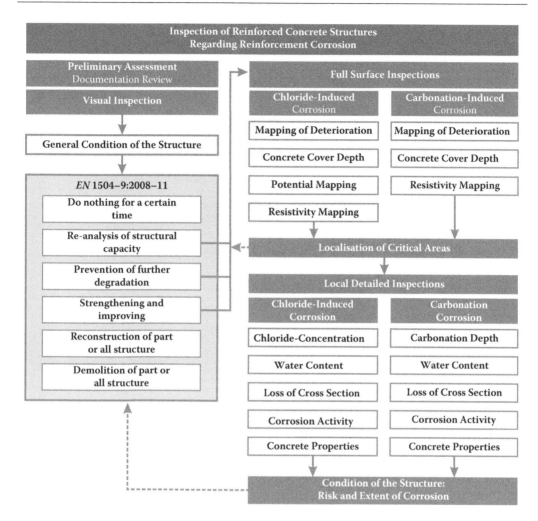

Figure 3.2 Procedure for the assessment of the condition of concrete structures regarding reinforcement corrosion according to Raupach et al. 2013.

structure contains such areas despite using boreholes, other nondestructive methods can be used to detect those areas. Most nondestructive methods are based on the measurement of acoustic or electromagnetic (radar) waves and are described in the following sections.

If both sides of the structure are accessible, the ultrasonic sound speed of the construction material can be measured as depicted in Figure 3.4. The test method is standardised in (DIN) EN 12504-4:2004, which not only describes different test setup possibilities, but also provides information on the influences of the environment, e.g., temperature, relative humidity (rh), and cracks within the specimen, as well as the dimensions of the samples of this specific test method.

Through the cross section of the structure pulsed ultrasonic waves are sent by the transmitter and recorded by the receiver unit placed on the opposite side of the wall. The wall thickness, which has to be known, as well as the time difference between the sending and receiving of the signal, can be used to calculate the ultrasonic speed, which is a material constant. This material constant varies between the typical construction materials and between different types of concrete, so that this value always has to be determined and cannot be given exactly in advance.

Figure 3.3 Example of very distinct gravel pockets at the surface of a concrete wall.

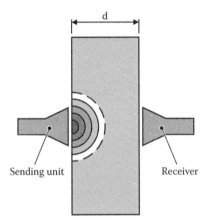

Figure 3.4 Schematic drawing of how to measure the ultrasonic speed of a building material.

The ultrasonic speed can be correlated to a certain extent with the compression strength of the concrete, while changes of the ultrasonic speed can indicate areas with hidden defects and voids. The results of the measurements should always be correlated with destructive test methods—regarding the compressive strength using additional compression tests on cores taken out of the structure, regarding voids and defects, e.g., with endoscopic investigations.

If the thickness of the concrete element and the time shift between sending and receiving the signal is known, the following formula can be used to calculate the ultrasonic speed of the concrete:

$$v_{sound} = \frac{d}{t}$$

where

v_{sound} = Ultrasonic speed in m/s
d = Thickness of the concrete element in m
t = Time shift between sending and receiving the signal in s

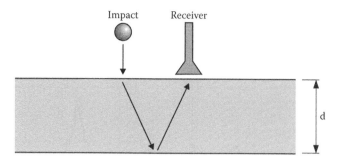

Figure 3.5 Schematic drawing of the principle of the impact-echo method.

As long as both sides of the structure are accessible, the previously shown method is suitable for on-site investigations. If both sides of the construction element are not accessible, but still voids or defects within the cross section have to be detected, the impact-echo method can be used as depicted in Figure 3.5. The impact-echo method was developed for measuring the thickness of an element, but can also be used to detect irregularities in an element.

The basic idea is nearly the same as mentioned before, but it is not an ultrasonic wave that is induced into the structure but a wave introduced mechanically, e.g., by a hammer. The induced sound waves are recorded by a receiver, which is usually mounted closely to the hammer. Because the sound wave is not reflected at a specific point, but the entire cross section around the measuring point, the receiver records a sound pattern. This sound pattern cannot be used for further analysis and has to be postprocessed. This is done by a fast Fourier transformation (FFT), which calculates a frequency spectrum (intensity over frequency) from the recorded sound waves. This allows the determination of the predominant frequency. This predominant frequency can then be used to calculate the thickness of the structural element according to the following formula:

$$v = 2 \cdot d \cdot f$$

where
   v = Speed in m/s
   d = Thickness of the concrete element in m
   f = Characteristic frequency of the multireflection in 1/s

Figure 3.6 shows the detection of voids by the impact-echo method on two different walls. The first example (a) shows a homogeneous wall section. With a known sound velocity of 3.42 km/s, the characteristic frequency of the multireflection is determined to be 3.42 kHz, and by using the given formula, the wall thickness can be calculated to 50 cm. The second example (b) shows the same wall with a void within the cross section. If the material is identical to the first wall, the sound velocity does not change, but the fast Fourier transformation calculates a characteristic frequency of 7.32 kHz, which leads to a thickness of 23 cm. This is equivalent to the depth of the void within the cross section.

Figure 3.7 shows an impact-echo device during operation on site. The different hammers are selected in dependency of the desired impact. The hammer itself is trigged manually, and the sound is recorded by the microphones in the silver cylinder. Not shown is the recording and processing unit.

Voids or irregularities resolve in a frequency shift in the resulting spectrum, as well as the predominant frequency, and thus indicate lower or greater thicknesses of the

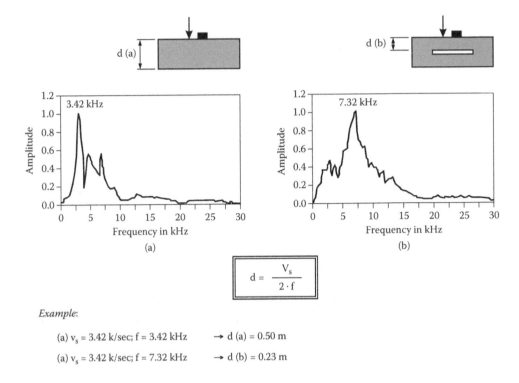

$$d = \frac{V_s}{2 \cdot f}$$

Example:

(a) v_s = 3.42 k/sec; f = 3.42 kHz   → d (a) = 0.50 m

(a) v_s = 3.42 k/sec; f = 7.32 kHz   → d (b) = 0.23 m

Figure 3.6 Impact-echo frequency analysis—example of two different measurement tasks.

Figure 3.7 Left: Impact-echo device placed on the concrete surface. Right: Impact-echo device during operation.

element, which then means that voids or defects are hidden in the structure. As mentioned before, a destructive verification of the test results is also recommended, e.g., by endoscopic investigations. (See Figure 3.8.)

Figures 3.9 and 3.10 were taken during the inspection of the mounting construction of concrete façade elements. The mounting elements are able to adjust the position of the façade element during mounting, as well as bear compressive and tension forces between the façade elements and the substructure.

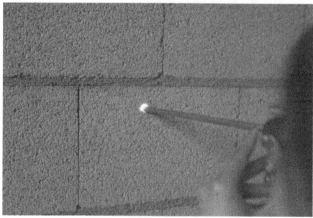

Figure 3.8 Inspections through a borehole with a rigid endoscope with fluorescent light source.

For large areas the previously mentioned methods of detecting hidden voids and defects will lead to an enormous amount of work, which can be very tedious, especially if ceilings have to be investigated. In this case as well as for large areas, radar, e.g., ground-penetrating radar (GPR), can be used.

Figure 3.11 shows a steel-reinforced arch below railway tracks. The shown areas without any concrete within the structure were detected accidentally. Due to the railway traffic there were no possibilities to perform time-consuming investigations of the arch from the upper side. So in addition to boreholes and endoscopic investigations from the lower side, GPR measurements were carried out from the upper side of the construction.

GPR was originally designed for geophysical measurements in order to detect buried objects or soil boundaries. It is also based on sending and receiving electromagnetic radiation, which are in the microwave band (UHF and VHF frequencies) and not introduced into the structure by a mechanical device but a complex sending unit.

Figures 3.12 to 3.16 show the basic principle, a handheld GPR device, as well as various results of different ground-penetrating radar measurements. The pictures clearly show that the thickness as well as the position of the reinforcement can be measured at once. Additionally, voids or changes in the concrete cover are also detected. Generally, GPR measurements produce results that have to be investigated carefully and require some experience.

Figure 3.9 Inspection of the mounting construction through joints of a concrete façade element—outside view.

Figure 3.14 shows a GPR recording of a defect-free concrete slab with a regular concrete cover as well as a constant thickness. It can be seen that the interpretation of this comparatively easy measurement task still requires some experience with GPR measurements because the readings themselves have to be interpreted by the operator.

Figures 3.15 and 3.16 show GPR measurements with a void as well as varying concrete cover. Like always, the measurements should be confirmed on a random basis with destructive tests.

Depending on the frequency used for GPR measurements, the maximum measurement depth as well as the resolution varies; see Table 3.1.

Numerous investigations regarding the advantages and disadvantages of the mentioned measuring techniques reveal that none of the different methods are capable of detecting all different types of possible defects or voids. Generally these investigations lead to the conclusion that the water content of the construction material mainly affects the reliability of the different measuring techniques. Comparatively, young and wet concrete can be best investigated with impact-echo or ultrasonic measurements. GPR measurements lead to reliable results, as long as the concrete is sufficiently dried out (Müller and Fenchel 2006).

As a conclusion for on-site investigations, it can comprehended that the various non-destructive methods can be combined in order to be able to detect voids or defects. Also, the nondestructive methods should always be calibrated and verified with destructive investigational methods, such as endoscopic investigations.

3.2.2.3 Compressive strength

The compressive strength of concrete is one of the key parameters regarding the load-bearing capacity of a concrete structure. It can fairly easily be determined on cores taken out of the structure, but can also be determined nondestructively. The judgement of the compressive strength of existing structures is standardised in (DIN) EN 13791:2008. This standard features different possibilities to determine the concrete strength of existing structures:

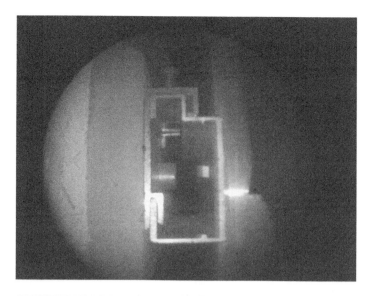

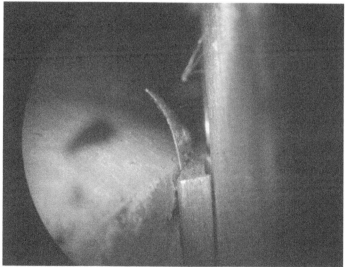

Figure 3.10 Inspection of the mounting construction through joints of a concrete façade element—view through the endoscope. Top: mounting detail. Bottom: detailed view of the mounting element located on the concrete.

1. Direct method
   a. Extraction of drill cores—test according to (DIN) EN 12504-1:2009
2. Indirect methods
   a. Rebound hammer method—test according to (DIN) EN 12504-2:2012
   b. Pullout force method—test according to (DIN) EN 12504-3:2005
   c. Ultrasonic pulse velocity—test according to (DIN) EN 12504-4:2004

If it is possible to take drill cores out of a structure, the compressive strength of these drill cores can be determined in an uniaxial compression test. The number of specimens taken

Figure 3.11 Example of a steel-reinforced railway bridge with missing concrete in between the upper parts of the arch.

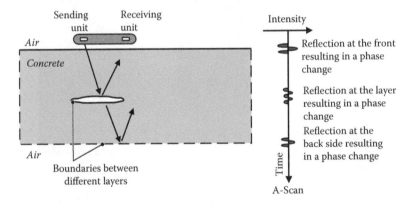

Figure 3.12 Schematic principle of the GPR measurement technique.

out of an existing structure should always be minimised. Also, the cores should not be taken in areas that are crucial to the load-bearing behaviour of the entire structure, such as cross sections close to a support or slender highly loaded columns. Also, in areas with prestressed steel cables the extraction of drill cores has to be carried out with extreme caution. The cores should be taken so that the reinforcement is not damaged or only damaged to an acceptable extent. If rebars are in the cores, they have to be located close to the end and also run perpendicular to the axis of the drill core. If rebars run parallel to the axis of the drill cores, the drill cores cannot be used to determine the compressive strength. Figures 3.17 shows an example of the extraction of drill cores out of a bridge deck, as well as one extracted core.

Depending on the number of drill cores extracted from the building or structural element, the characteristic compressive strength is determined with two different procedures

Figure 3.13 Handheld GPR device. Left: GPR during operation. Right: Results during measurement.

according to (DIN) EN 13791:2008. If the number of extracted drill cores is greater than 15, the following formula can be used to determine the characteristic compressive strength:

$$f_{ck,is} = f_{m(n),is} - k_2 \cdot s \text{ or } f_{ck,is} = f_{is,lowest} + 4$$

where

$f_{ck,is}$ = Characteristic compressive strength of the concrete.
$f_{m(n),is}$ = Arithmetic mean of n test results of the compressive strength of the concrete.
$f_{is,lowest}$ = Lowest test result of the compressive strength of the concrete.
k_2 = Defined in national standards. If they don't exist, $k_2 = 1.48$ is assumed.
s = The standard deviation of the test results or 2 N/mm². The greatest value is to be used.

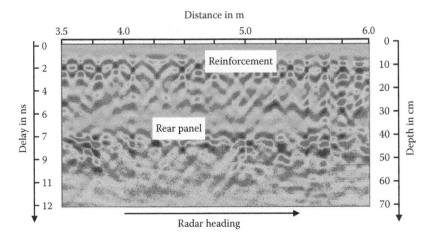

Figure 3.14 GPR measurement of a defect-free concrete slab.

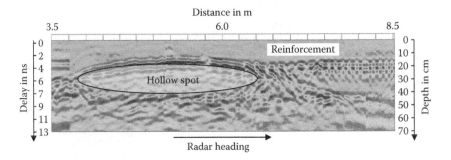

Figure 3.15 GPR measurement showing a concrete slab containing a void as well as reinforcement.

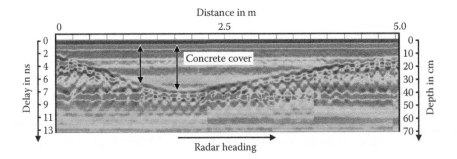

Figure 3.16 GPR measurement showing a concrete slab with changing concrete cover.

Table 3.1 Frequency, maximum depth, and resolution of GPR

Frequency MHz	Maximum depth cm	Resolution	Typical application
1	2	3	4
250	700		Geotechnics, location of underground objects —
500	300		
1000	100		Detection of
2000	30		rebars and voids

Figure 3.17 Top: Extraction of drill cores out of a bridge deck. Bottom: Drill core taken out of a bridge deck.

If the number of drill cores is between 3 and 14, the following formula is used to determine the characteristic compressive strength:

$$f_{ck,is} = f_{m(n),is} - k \text{ or } f_{ck,is} = f_{is,lowest} + 4$$

The range of k for a small number n of test results is defined as follows:

n	k
10–14	5
7–9	6
3–6	7

The characteristic compressive strength can then be correlated with a concrete class according to Table 3.2. It has to be noted that the following is part of the national annex of (DIN) EN 13791:2008 and may vary throughout the national standards.

Depending on national test procedures, the dimensions of the drill cores might vary. According to (DIN) EN 13791:2008 as well as (DIN) EN 12504-1:2009, the diameter of a single drill core should to be at least three times the size of the maximum aggregate size. In Europe, usually the maximum diameter of aggregates in large concrete elements varies between 16 and 32 mm. If no documents of the concrete composition are available, a core should have a diameter of 100 mm. As mentioned above, generally specimens with reinforcing bars in longitudinal directions cannot be used for determining the compressive strength of concrete.

If it is not possible to extract a large number of drill cores out of the structure, the previously mentioned indirect methods enable us to limit the number of drill cores and

Table 3.2 Class of compressive strength vs. characteristic minimum compressive strength of concrete (determined on samples taken on site)

Class of compressive strength according to (DIN) EN 206-1	Ratio of the compressive strength of concrete to the characteristic compressive strength of standardised samples	Characteristic minimum compressive strength of concrete N/mm²	
		$f_{ck,is,cylinder}$	$f_{ck,is,cube}$
C8/10	0.85	7	9
C12/15	0.85	10	13
C16/20	0.85	14	17
C20/25	0.85	17	21
C25/30	0.85	21	26
C30/37	0.85	26	31
C35/45	0.85	30	38
C40/50	0.85	34	43
C45/55	0.85	38	47
C50/60	0.85	43	51
C55/67	0.85	47	57
C60/75	0.85	51	64
C70/85	0.85	60	72
C80/95	0.85	68	81
C90/105	0.85	77	89
C100/115	0.85	85	98

might allow the determination of the compressive strength in large areas nondestructively (rebound hammer and ultrasonic pulse velocity). None of the indirect methods determine a compressive strength directly but determine a specific value (depending on the selected test method), which can then be correlated to the compressive strength of a small amount of drill cores. The correlation between the results of the direct and indirect methods is also given in (DIN) EN 13791:2008.

The most common way is to use a rebound hammer (also called Schmitthammer) (see Figure 3.18). The pictures show the schematics of the rebound hammer as well as the rebound hammer during operation of a concrete surface.

The execution of tests with this rebound hammer is regulated in (DIN) EN 12504-2:2012. The basic idea behind this test is to measure the rebound of an impact of a defined mass off the concrete surface. Generally the minimum thickness of the building part has to be 100 mm, and the testing area has to be at least 300×300 mm^2. Each area has to be tested with nine single strikes. The median value then represents the rebound value for this specific test area.

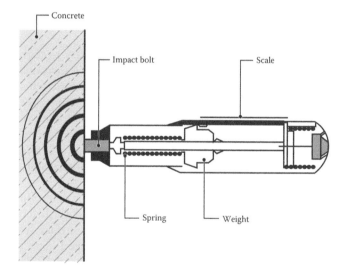

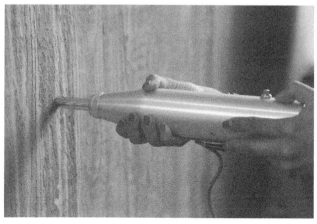

Figure 3.18 Top: Schematic drawing of the rebound hammer. Bottom: Rebound hammer during operation.

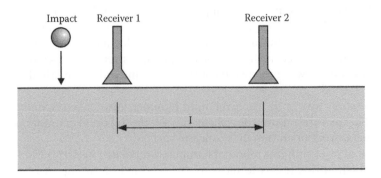

Figure 3.19 Schematic drawing of how to measure the surface velocity with impact echo.

It has to be considered that the rebound value is affected by the condition of the surface, the carbonation depth, the age of the concrete, as well as the applied load. In order to derive a compressive strength out of the rebound values, (DIN) EN 13791:2008 suggests different methods. The most reliable one is to correlate the rebound values with the compressive strength determined on drill cores taken within a limited number of test areas. The rebound hammer enables the engineer to enlarge the number of test areas in comparison to the extraction of drill cores.

The standard (DIN) EN 13791:2008 also gives the possibility to correlate the rebound values directly with compressive strengths of concrete. Our own investigations clearly indicate that this correlation should not be used under any circumstances because it cannot be stated whether the compressive strength is under- or overestimated.

Another destructive method is the determination of the pullout force of metal bolts embedded into the concrete using drill holes. This method is further described in (DIN) EN 12504-3:2005. In order to provoke a concrete failure during the test, a so-called undercut dowel has to be used. The correlation of the pullout forces with the compression strength has to be done according to (DIN) EN 13791:2008.

The compressive strength can also be determined indirectly by determination of the ultrasonic pulse velocity according to (DIN) EN 12504-4:2004, as described in Section 3.2.2.2. Another possibility of the test setup is the determination of the surface velocity of the material. The surface velocity is measured with the previously described impact-echo technology, but two (instead of one) receiving units have to be used. See Figure 3.19. The surface velocity is then calculated by using the distance of the two recording units as well as the time difference of the sound events recorded with each unit.

$$v_{sur} = \frac{l}{\Delta t}$$

where

v_{sur} = Surface velocity in m/s
l = Distance of the two recording units in m
Δt = Time difference of the sound events recorded with each unit in s

3.2.2.4 Surface strength

The surface strength is one of the most important material properties concerning existing concrete structures. It can deliver qualified information about the quality of the border zone of the concrete structure.

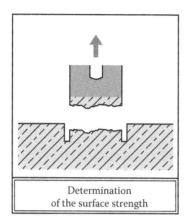
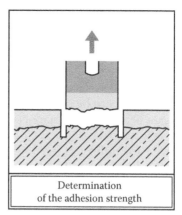

| Determination of the surface strength | Determination of the adhesion strength |

Figure 3.20 Left: Determination of the surface strength. Right: Determination of the adhesion strength.

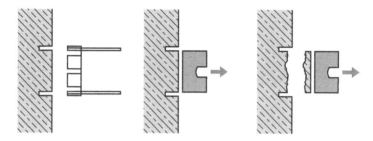

Figure 3.21 Determining the surface strength according to (DIN) EN 1542:1999. (From Momber, A.W., Schulz, R.-R., *Handbuch der Oberflächenbearbeitung Beton*, Basel: Birkhäuser, 2006.)

Figure 3.20 illustrates the difference between the surface strength of concrete and the adhesion strength of a surface protection system or a mortar-based coating on top of concrete or masonry. Generally speaking, if the rupture occurs in the substrate, the adhesion strength of the coating to the concrete is greater than the surface strength of the substrate (concrete), and in this case the test delivers the surface and not the adhesion strength. Despite the different types of strength, the test itself is conducted equally.

Figure 3.21 illustrates the basic test procedure according to (DIN) EN 1542:1999, which regulates the determination of the surface strength of concrete. This standard is very comprehensive and covers all possible cases, which can occur while investigating an existing structure.

The test is usually conducted on site or in the laboratory on smaller samples in the following manner:

- Drilling of a cylindrical nut with a diameter of 50 mm and a depth of at least 10 mm into the substrate
- Applying a steel die with a reactive adhesive, e.g., (sand-filled) epoxy resin
- Cleaning of the nut so that no adhesive is in the groove and falsifies the result
- Adapting of the test rig to the steel die and performing an uniaxial tension test with a load rate of 100 N/s (concrete surfaces or stiff surface coatings) or 300 N/s (elastic surface coatings)
- Calculating of the tensile strength as well as judging the type of rupture in percentiles of the entire cross section of the die

The judging of the rupture percentiles is usually done based on experience of the test personnel and does not include complex measurements. The type of rupture refers to the location of rupture within the material, e.g.:

- X% substrate
- X% cohesion in the surface coating (if more than one layer is applied on the surface, the type of layer has to be indicated)
- X% adhesion between the surface coating and the substrate

In order to avoid any influences due to the location, the rules given in Figure 3.22 should be observed. (DIN) EN 1542:1999 defines 50 mm as the diameter of the steel die; thus, all mentioned dimensions in this standard are only valid for this diameter. If the diameter of the steel die has to be adapted, the given rules enable us to calculate individual dimensions.

Besides the location of the test areas, the steel dies that are used to adapt the test rig onto the surface of the structure should have certain dimensions. These dimensions are required

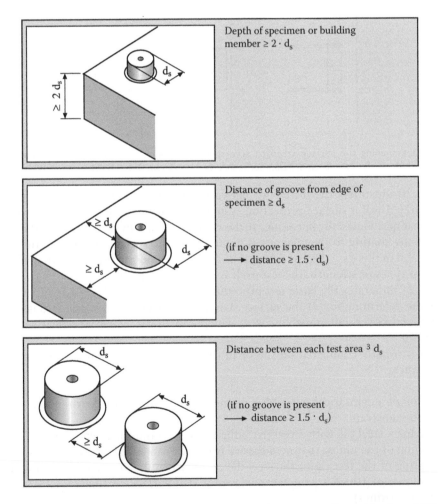

Figure 3.22 Requirements for the selection of the position of test areas. (From Momber, A.W., Schulz, R.-R., *Handbuch der Oberflächenbearbeitung Beton*, Basel: Birkhäuser, 2006.)

Figure 3.23 Mobile test rig used to determine the surface as well as the adhesion strength.

so that the die does not wrap due to the test load. Additionally, it is also required to drill at least 10 mm into the substrate. The drilling ensures that the tension is applied unidirectionally in the surface, and the rupture only occurs in a defined area.

The surface strength of a concrete structure as well as the adhesion strength (Figure 3.23) is determined in different states of repair and retrofitting (see Chapter 7):

1. Determination of the current condition
2. After the preparation of the surface and before the application of any surface protection system
3. After application of any surface protection system

3.2.3 Concrete cover

3.2.3.1 General

Detailed information about the location and type of the embedded reinforcement is a key parameter regarding any analysis of the load-bearing behaviour. Regarding corrosion protection, the location of the reinforcement is also a key parameter, which has to be known because adequate cover depths are essential to ensure a sufficient durability of a construction. The minimum cover depths are regulated by Eurocode EC 2 and (DIN) EN 206-1:2001, depending on the exposure of the construction.

Usually engineers need to know the concrete cover as well as the type of reinforcement of an entire structure and not only at a limited number of spots. This can only be achieved by nondestructive methods. Depending on the intention of the investigation, these methods also have to be calibrated by local inspections.

Figure 3.24 shows an example of a basement garage in which the concrete cover varies between 0 and 30 mm, and thus is too low in some areas for this specific exposure. These areas can only be detected by an extensive investigation and not by a local analysis of randomly selected points.

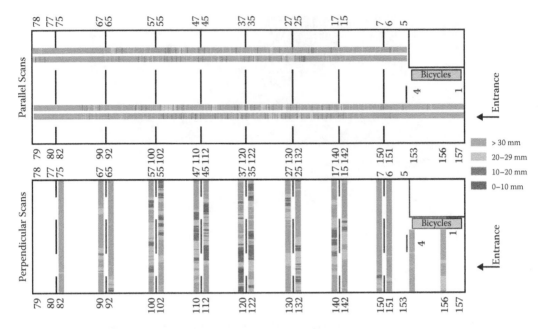

Figure 3.24 Determining the concrete cover—results of line scans in a basement garage.

Generally steel reinforcement can be detected nondestructively by using magnetic or inductive methods. As shown in Section 3.2.2.2, GPR can also be used to detect the reinforcement, but due to the complexity of GPR measurements, this method is not commonly used to detect steel rebars.

3.2.3.2 Magnetic methods

Magnetic methods can only be used for a rough estimation of the concrete cover as well as the location of the reinforcement as long as the rebars are located close to the surface. The method is based on the fact that regular steel reinforcement can be magnetised by an externally applied magnet.

The easiest method is to scan the concrete surface with a magnet and measure the force induced by the magnetic attraction of the magnet and the rebars. The closer the magnet is to a rebar, the higher the force due to the magnetic attraction. If the test setup is calibrated, the force corresponds to the distance between the magnet and the rebars, and thus the concrete cover can be calculated.

Besides the force induced by the magnetic attraction, the magnetic field of the surface magnet can be measured. This magnetic field remains constant as long no rebar or other metal object is located in this magnetic field. If a rebar is located in the field, the field is altered, which can be detected. This change of the magnetic field can also be calibrated and correlated with a concrete cover.

3.2.3.3 Inductive methods

Measuring the concrete cover by measuring the inductivity can also be done in two different setups. The basic principle of the two different test methods is identical and depicted in Figure 3.25. A yoke magnet carrying an exciter and receiving coil is placed on top of the construction and the magnetic flux is measured.

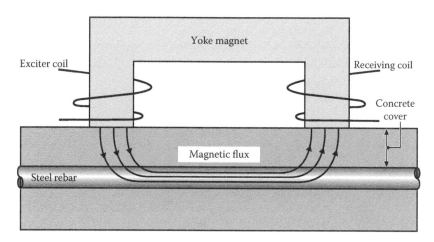

Figure 3.25 Measurement principle of the inductive method to determine the concrete cover nondestructively.

The alternating field method works like a power transformer. An alternating current power supply is connected to the exciter coil, which induces an alternating current in the receiving coil. The current induced into the receiving coil is influenced by any magnetic material close to the yoke magnet, and so steel rebars that might be in the magnetic field affect the induced current. The current is influenced by the distance between the rebar and the yoke magnet as well as the diameter of the rebar. After calibrating the test setup with no additional magnetic material in the vicinity of the yoke magnet, the concrete cover or the diameter of the rebar can be measured. If the diameter of the rebar is known, the concrete cover can be measured or vice versa. One of the two variables has to be detected destructively and cannot be determined nondestructively.

The second method is the so-called eddy current method, which enables measurements without the receiving coil. The exciter coil generates an alternating magnetic field and induces an eddy current into the rebars. The induced eddy current also generates a magnetic field that superposes the magnetic field in the exciter coil. This superposition leads to changes of the impedance of the exciter coil, which enables us to determine the concrete cover and the diameter of the reinforcement independently. Additionally, this method can be used if nonmagnetic reinforcement, e.g., stainless steel, has been placed in the construction. Figure 3.26 shows a rebar analyser during operation.

3.2.4 Position and diameter of the reinforcement

In order to determine the cover depth as well as the position of the reinforcement, often so-called line scans are conducted. The sensor is moved along a linear scan path. Because the sensors can only detect changes in a magnetic field, only rebars that are placed rectangular to the path can be detected. Figure 3.24 is thus based on 21 single scans. Line scans are not feasible to measure the diameter of the rebars in addition to the position and the cover depth. If the diameter has to be detected additionally, so-called area scans have to be conducted.

Figure 3.27 shows the results of a line scan (bottom) and an area scan (top). The area scan is obtained by eight single line scans, which are rectangular to each other and cover an area of 60 × 60 cm². Depending on the system used, the diameter as well as the concrete cover is calculated after completing the area scan.

Figure 3.26 Rebar analyser during operation. Top: Scanning unit. Bottom: Scanning and processing unit.

Because the position of the single line scans is crucial to the result, references lines should be used. Also, prefabricated patterns are available (see Figure 3.28).

If the exact rebar cross section has to be known, the previously mentioned methods are not resilient and inspection holes have to be made. One possible method is to use high-pressure water jets with a working pressure up to approximately 100 MPa. The opening of a construction by using high-pressurised water has the advantage that the reinforcement is not affected as well as damaged by the concrete removal, and corrosion products are also removed. Additionally, the concrete can also be removed behind the reinforcement bars, and thus the entire rebar can be investigated.

So the diameter as well as the remaining cross section of the reinforcement can be investigated at once. Figures 3.29 and 3.30 shows the opening of a bridge beam by the high-pressure water jet method, as well as the rebars after removing the concrete.

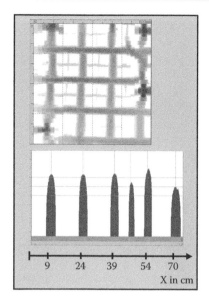

Figure 3.27 Results of an area scan (top) and a line scan (bottom) showing the location of the rebars.

Figure 3.28 Prefabricated reusable pattern to conduct area scans on site.

After finishing the survey of the reinforcement, the openings have to be closed properly. In order not to damage the structure by the opening, the locations of these openings should always be determined in cooperation with a structural engineer.

The dimensions of the rebar can be determined either on site by using a calliper gauge (see Figure 3.31) or extracting parts of the reinforcement or in the lab. If the reinforcement is extracted from the structure, a structural engineer has to be consulted before any action. While measuring on site, it is important that all sides of the rebars are accessible and hidden corrosion pits are not neglected.

If an extraction of a part of the reinforcement is possible, the rebars can be cleaned with a special acid in a laboratory and then measured by using a microscope or special scanners (see Section 4.3.3.4).

Figure 3.29 Removing the concrete cover with high-pressure water jetting.

Figure 3.30 Reinforcement after removing the concrete cover.

3.2.5 Cracks

3.2.5.1 General

Cracks in constructions occur due to many different reasons. Table 3.3 gives an overview of the most common cracks in concrete structures following the German concrete association (DBV 2006).

Figure 3.32 illustrates cracks running along the reinforcement in straight lines. This crack pattern is typical for working joints or joints between prefabricated concrete parts covered by a thin layer of concrete.

Figure 3.31 Determination of the diameter of the reinforcement after removing the concrete cover by using a calliper gauge.

Besides the different types of cracks and their origin, cracks can contain different amounts of water. Figures 3.33 to 3.36 show typical cracks in concrete structures, beginning with a dry crack and ending with a crack leading pressurised water.

3.2.5.2 Crack mapping

The mapping of cracks is usually done for an entire structure by visual inspection. It includes the location, course, and crack width, which are all transferred into plans or sketches of the structure. The crack mapping is usually done together with the mapping of defects and voids. For a better detection of small cracks, the concrete surface can be prewetted with an aerosol can. Immediately after the surface is nearly dried, small cracks are highlighted and can be detected comparatively easily. This also works for rough surfaces, such as sprayed concrete surfaces. (See Figure 3.37.)

Depending on the type of structure, a grid system may be introduced before detecting the cracks. A grid system allows a fairly easy positing of each crack. This grid system can be oriented on structural elements, such as regular columns in a parking deck. Additional photos of the entire structure, taken during the crack mapping, can also bear valuable information for an additional investigation in the office.

Figure 3.38 shows regular and parallel cracks running along the edges of beams below the ceiling. The origin of these cracks is due to an abrupt change of the stiffness of the entire construction, and thus drying shrinkage of the entire structure lead to cracks at the area where the stiffness changes.

The crack width can be measured by using loupes with enclosed scales or by a crack width ruler, as shown in Figure 3.39. The crack width ruler contains black marks, usually beginning with 0.1 mm and ending with 5 mm. By a comparison of the crack width as well as the marks on the ruler, the crack width can be determined. It has to be noted that the crack width directly on the surface might be larger than the crack width in most parts of the crack, so the drilling of a core through the crack might be necessary to measure the correct crack width (see Section 3.2.5.3).

Table 3.3 Different types of cracks: causes, appearance, and short description

Cause of cracks	Appearance	Description
Cracks due to properties of the concrete		
Crazing	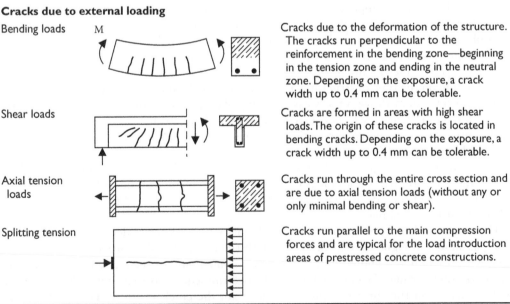	Crazing is fine cracking close to the surface due to shrinkage of the surface layer of concrete, e.g., due to a lack of a sufficient curing. The cracks are usually only a few millimetres deep.
Drying shrinkage		Cracks due to the reduction of the entire concrete's volume caused by shrinkage (of the entire cross section) in combination with a hindered structure, e.g., large slab. Usually the cracks run through the entire cross section.
Voids below the reinforcing bars		Cracks run parallel to rebars close to the surface.
Cracks due to external loading		
Bending loads	M	Cracks due to the deformation of the structure. The cracks run perpendicular to the reinforcement in the bending zone—beginning in the tension zone and ending in the neutral zone. Depending on the exposure, a crack width up to 0.4 mm can be tolerable.
Shear loads		Cracks are formed in areas with high shear loads. The origin of these cracks is located in bending cracks. Depending on the exposure, a crack width up to 0.4 mm can be tolerable.
Axial tension loads		Cracks run through the entire cross section and are due to axial tension loads (without any or only minimal bending or shear).
Splitting tension		Cracks run parallel to the main compression forces and are typical for the load introduction areas of prestressed concrete constructions.

Source: Deutscher Beton-Verein (DBV), DBV Merkblattsammlung, *Merkblatt Rissbildung: Begrenzung der Rißbildung im Stahlbeton- und Spannbetonbau*, Wiesbaden: Deutscher Beton- und Bautechnik-Verein e.V., 2006.

Figure 3.32 Example of cracks following straight lines that are typical for working joints.

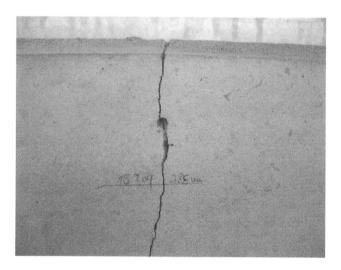

Figure 3.33 Example of a wide dry crack.

The choice of an accurate crack treatment is strongly influenced by the crack width, the movement of the crack, as well as the condition of a crack, e.g., if the crack bears water or if the crack width changes over time. Therefore, this information should always be documented during a crack mapping.

Figure 3.40 shows a crack that contains water and is already self-healed to a certain extent. So a nonpressurised crack injection cannot be conducted because the injection material would not penetrate the crack. Not only self-healing of cracks but also preliminary crack treatments can play a major role concerning the success of a crack treatment, and thus they also have to be recorded.

Figure 3.41 shows a typical deterioration map (Reichling et al. 2013) in which cracks, wet areas, spalling, and delamination are recorded.

Figure 3.34 Example of a moist crack with lime marks.

Figure 3.35 Example of a moist crack and depressurised water running through the crack.

Figure 3.36 Example of a wet crack and pressurised water running through the crack.

Figure 3.37 Crack mapping on a rough concrete surface. Top: Dry crack. Bottom: Clearly visible crack due to wetting of the surface.

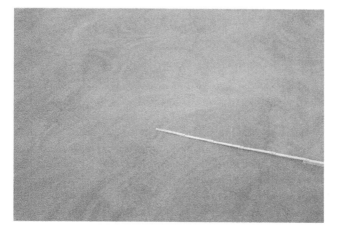

Figure 3.38 Example of parallel cracks that were detected on top of beams holding the depicted slab.

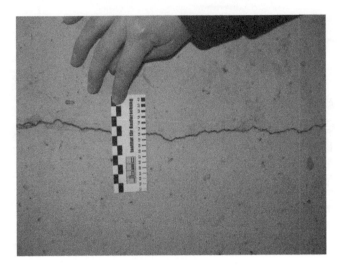

Figure 3.39 Measuring the crack width with a special ruler.

Figure 3.40 Example of a partially closed crack due to self-healing.

3.2.5.3 Crack depth

Regarding the crack depth, it is important to know if the crack extends over the entire thickness of the building member (separation cracks). This can easily be accomplished by comparing both sides of the building member. The determination of the crack depths (e.g., structural elements with one-sided accessibility) can be done in a destructive way (e.g., by sounding drillings) (see Figure 3.42) or by using nondestructive methods (e.g., by impact echo, ultrasonic measurements). Due to the time-consuming investigation methods, these methods are usually limited to local investigations.

The crack depth may also be determined by injecting a low-viscose polymeric resin into a crack and measuring the crack depth after the resin has hardened and a drill core has been taken directly over the crack. The resin ensures that the drill core is not separated due to the crack, and additionally the crack width can be detected on the drill core.

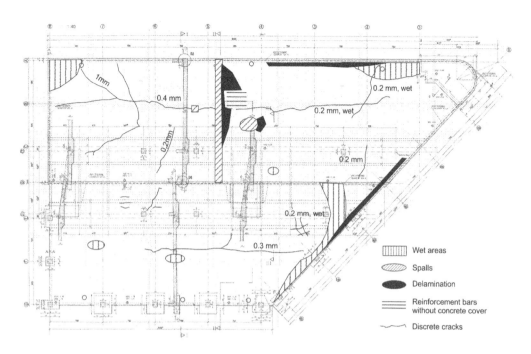

⠿⠿⠿	Wet areas
⊘	Spalls
●	Delamination
≡	Reinforcement bars without concrete cover
⌇	Discrete cracks

Figure 3.41 Example of a deterioration map of the ceiling in a car park. (From European Federation of Corrosion Working Party 11, Corrosion of Steel in Concrete, *Materials and Corrosion* 64 (2013), no. 2.)

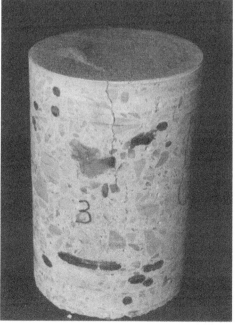

Figure 3.42 Left: Drill core with separating crack. Right: Drill core with a crack ending in the middle.

3.2.5.4 Crack movements

The knowledge of crack movements is important because depending on possible movements, the types of crack repair method and injection material have to be selected. There are different methods to detect crack movements.

Gypsum or mortar marks are cheap and easy to use for the investigation of long-term movements. The marks are applied perpendicular to the crack, where a crack opening results in a crack in the gypsum mark. The crack edges in the gypsum or mortar allow a precise measurement of the opening width. Cement-based mortars are often used for the application on external surfaces.

Alternatively, prefabricated marks can be used. These marks are fixed to the surface along the crack edges by using a fast adhesive resin. Scales on these marks enable the recording of crack movements. Other common methods for recording the long-term crack movements are linear variable differential transformers (LVDTs), which allow a permanent recording of the crack movement. Figure 3.43 shows parts of a crack monitoring system installed on a roof of a concrete structure. The measurement is conducted with LVDTs. Also, the system includes a recording unit, which is able to send the data via mobile services, e.g., to an email account. This allows receiving the data without accessing the recording unit. Also, the temperature is recorded. If monitoring crack movements, the temperature should always be recorded in order to interpret them.

3.2.6 Danger of concrete corrosion

3.2.6.1 General

The following describes analytical methods to determine which reasons and mechanisms lead to damage of the concrete. The basic mechanisms that lead to a deterioration are described in Section 2.21.

3.2.6.2 Acid attack

Concrete can be damaged by acid solutions, such as sulphuric acids, which is the most common acid solution damaging concrete. The acid attack of concrete can occur due to the contact with sulphur-containing soil or due to an acid atmosphere in sewer pipes (biogenic sulphuric acid corrosion).

In order to correlate the depth of damaging with an acid concentration, concrete samples have to be taken out of the structure, and visual as well as chemical investigations have to be conducted. The visual investigations are usually done on epoxy-impregnated and cut concrete samples in order to determine the depth of damaging visually. An example of such a specimen is shown in Figure 3.44. In the image the damage due to an acid attack can be clearly visualised and the depth can be determined. In order to assess the critical depth chemically, concrete powder is usually extracted from the concrete and analysed afterwards.

The extraction of the concrete powder can be done analogue to the extraction of powder used for determining the chloride content of the concrete (see Section 3.2.7.3). It is important that the powder is obtained without using water because water will leach out sulphur, and thus the analysis will be impaired.

After extracting concrete powder the samples are dried at 105°C and the powder is analysed chemically. If the amount of sulphate has to be determined, e.g., a carbon/sulphur combustion analyser or closed acid digestion is used.

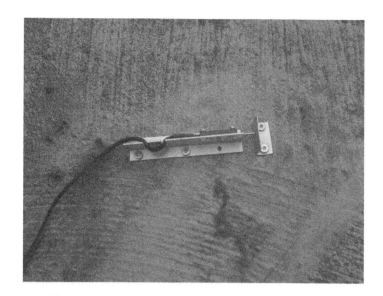

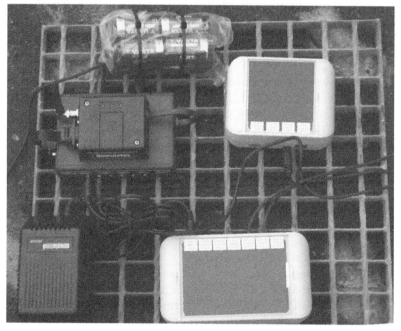

Figure 3.43 Top: LVDT crossing a crack. Bottom: Measurement, recording, and sending unit for measuring crack movements.

3.2.6.3 Freeze–thaw attack

Concrete that might be subject to freeze-thaw attack is usually analysed by optical methods. Freeze-thaw attack leads to distinct microcracks within the border zone of a building element. These cracks can be visualised by fine-cut microscopy on samples taken out of the structure, e.g., drill cores.

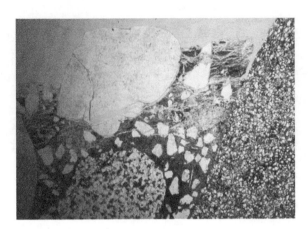

Figure 3.44 Detailed view of concrete damaged due to acid attack (specimen embedded in epoxy resin and cut afterwards).

Figure 3.45 shows typical pop-outs on a concrete surface subjected to a freeze-thaw attack, which was verified by fine-cut microscopy in order to rule out any other attack, e.g., mechanical attack.

3.2.6.4 Alkali–silica reaction

The determination of whether concrete is damaged by alkali-silica reaction is generally done on drill cores taken out of the damaged structure. The drill cores are then investigated by either petrographic analysis or SEM. The petrographic analysis allows the visualisation of the cracks formed around as well as through the aggregates, and the gel produced during the alkali-silica reaction. The SEM analysis enables the identification of the gel by using x-ray microanalysis.

Figure 3.46 shows concrete severely damaged by an alkali-silica reaction (ASR). Drill cores were taken out of the concrete and investigated by petrographic analysis.

Figure 3.47 shows sections of concrete embedded in a fluorescent epoxy resin and investigated by petrographic analysis. The concrete was severely damaged by ASR, which can be seen by the crack running through the aggregates (left) as well as the gel formed around the aggregates (right).

If additional investigations are necessary, drill cores can be tested according to national guidelines that deal with the avoidance of alkali-silica reactions in concrete. Usually concrete beams or cores are stored under elevated temperature, e.g., 40°C, and in a humid environment in order to provoke an alkali-silica reaction that can be determined by a significant elongation of the cores. This test can also be done with nondamaged drill cores extracted out of an existing structure to determine the elongation capacity of the concrete by ASR.

3.2.6.5 Identification of minerals

The identification of minerals can be done by using so-called scanning electron microscopy (SEM), which allows the visual analysis of broken or polished surfaces with large magnifications. The image is produced by an electron beam that is focused on the specimen. The electrons interact with the electrons of the material, and thus different signals are produced and recorded with different detectors. The scanning of the surface is generally done in a raster scan pattern. The detected signals as well as the exact position of the electron beam

Figure 3.45 Pop-outs on a concrete surface subjected to a freeze-thaw attack. (Photos taken by Consulting
        Engineers Raupach, Bruns, Wolff.)

at all times of the investigation allow the production of an image of the topography of
the samples.

Depending on the type of microscope used, the investigations are done in high vacuum,
low vacuum, or under environmental conditions (ESEM), which allows the investigation of
wet samples. For all other SEM analysis, the samples have to be dried before.

During the scanning process x-rays also occur. The energy of the x-rays depends on the
atomic number of the emitting atom, and thus can be correlated with the chemical element.
This allows a quantitative and qualitative determination of elements during a SEM analysis.

Figure 3.48 shows a polished cross section of concrete with a textile (AR glass) reinforcement embedded into it. In order to receive such images, the sample has to be dried and completely filled with, e.g., an epoxy resin, which enables us to polish the surface and thus create

Figure 3.46 Concrete severely damaged by ASR.

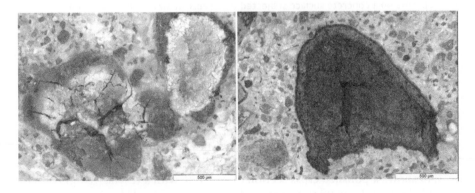

Figure 3.47 Petrographic analysis of severely damaged aggregates by ASR with cracked aggregates as well as gel forming around the aggregates.

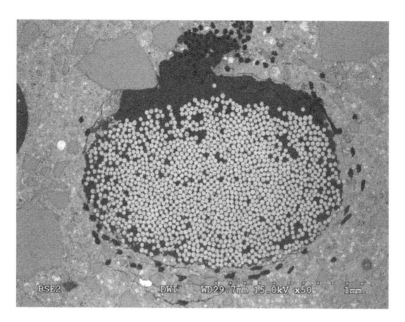

Figure 3.48 Textile reinforcement embedded in concrete. (Picture from DWI/RWTH Aachen University.)

a smooth and plane surface. The picture in Figure 3.48 was made at the DWI at RWTH Aachen University.

3.2.7 Danger of reinforcement corrosion

3.2.7.1 General

The following section describes analytical methods to determine what led to corrosion of the reinforcement. The description of the basic deterioration mechanisms can be found in Section 2.3.

3.2.7.2 Depth of carbonation

The depth of carbonation can be determined by various investigation methods, such as:

- Phenolphthalein indicator test
- X-ray diffraction analysis
- Infrared spectroscopy
- Differential scanning calorimetry
- Chemical analysis

The most common way to determine the depth of carbonation is the phenolphthalein indicator test method, which is regulated in (DIN) EN 14630:2006. The test method is based on the fact that phenolphthalein solution does change its colour from colourless to pink above pH 9. Usually the indicating solution contains 1 wt% phenolphthalein solved in alcohol (70 vol%). This indicating fluid is sprayed on a freshly broken concrete surface. The required amount of indication fluid varies due to the concrete porosity, so an absolute value cannot be given. The amount of solution should be sufficient for an intensive pink colour, which is automatically received on freshly broken noncarbonated concrete. The carbonated concrete remains colourless (see Figure 3.49).

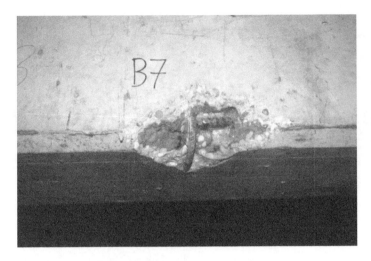

Figure 3.49 On-site determination of the depth of carbonation.

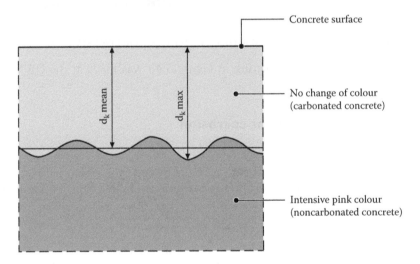

Figure 3.50 Schematic drawing of mean and maximum depth of carbonation determined with an indicator fluid (phenolphthalein solution).

The depth of carbonation can be measured after the solvent is evaporated and the surface dried off with a calliper gauge. Depending on the topography, it might be necessary to use an additional level to straighten out the surface. It is important that the indicating fluid is applied to broken surfaces and not to cut surfaces. The cutting dust does alter the measurements and produces misleading results.

As indicated schematically in Figure 3.50, the depth of carbonation (d_K) varies, and thus not only the mean value, but both the mean and the maximum value should be determined (see also (DIN) EN 14630:2006).

Along cracks of voids so-called carbonation cavities (see Figure 3.51) can occur. If the maximum carbonation depth in such a cavity is much greater than the mean value of carbonation depth, the depth of the carbonation cavity is not included in the calculation of the mean carbonation depth (see (DIN) EN 14630:2006). But the depth of the carbonation cavity should be recorded and preferably sketched or photographed.

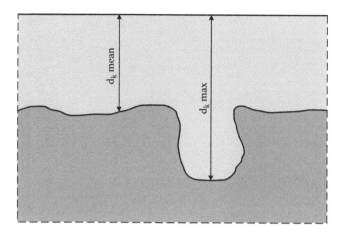

Figure 3.51 Schematic drawing of a carbonation cavity determined with an indicator fluid (phenolphthalein solution).

If a determination on site is not possible, drill cores with a diameter between 50 and 100 mm can also be used to determine the carbonation depth. After the extraction of the drill cores and before the laboratory investigations, the drill cores should be stored under the exclusion of CO_2, e.g., in closed plastic bags. If only a minimum of drill cores can be taken during a full survey, the determination of the depth of carbonation can be combined with the determination of the compressive strength. After determining the compressive strength on drill cores, the depth of carbonation can be determined on the remaining halves of the cores. This combination of two investigations bears the risk that the cores do not break into two halves, but into small pieces, which does not allow a proper measurement of the depth of carbonation.

3.2.7.3 Chloride profiles

The determination of the chloride distribution in a structure is not as simple as determining the depth of carbonation, because not only the presence of chloride has to be known, but also the concentration of chloride within the concrete cover. The concentration of chlorides in concrete can be determined by various chemical analyses of concrete powder (see below). The extraction as well as the basic analysis are regulated in (DIN) EN 14629:2007.

The excavation of concrete powder can basically be done in two different ways. Concrete cores can be taken out of the structure and then cut into small slices and milled. The cutting and milling process can be combined by a computer-controlled grinding machine, which is shown in Figure 3.52. This grinding machine was developed at ibac and allows the grinding of the concrete in defined depth intervals, and thus a very accurate determination of the chloride content in dependency of the position within the cross section. In order to avoid an influence of washing out the chlorides by cooling water during the grinding process, the presented grinding machine uses no water at all. Whether the water during the drill core extraction affects the total amount of chlorides depends on many boundary conditions and cannot be stated absolutely.

Another possibility is the excavation of concrete powder directly with a drilling machine, as shown in Figure 3.53. This drilling machine is equipped with special hollow drills that convey the concrete powder to a cyclone and also does not use any water at all. The cyclone conveys the concrete powder to a small container. The minimum amount of concrete

Figure 3.52 Top: Extraction of drill cores. Bottom: Computer-controlled grinding machine for milling concrete in defined depth intervals.

powder necessary for the investigations is 1 g for each depth. Usually the investigated depth is selected according to the concrete cover. Common depths are between 45 and 60 mm in steps of 10 or 15 mm. In order to achieve representative specimens, the drilling should be done in more than one spot.

After the excavation of the concrete powder and before any further analysis, the powder must be transported in lockable vessels or bags where the samples are protected against pollution.

The chemical analysis is usually carried out by potentiometric titration, but also other methods may be used, e.g., direct potentiometry or photometry. Regarding the potentiometric

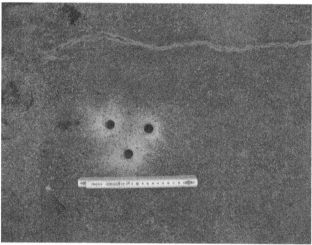

Figure 3.53 Top: Drilling machine with an integrated system to collect the powder samples. Bottom: Distribution of boreholes for determining the chloride content of concrete.

titration the samples must be dried and ground before the analysis. Also, a repeated determination is recommended.

In order to determine the chloride content, the powder sample is typically digested with nitric acid to dissolve all chlorides. Furthermore, ions (e.g., carbonates, sulphides) are eliminated that would influence the chemical analysis. The solids are filtered out and the chloride-containing solution is used for the analysis. Silver nitrate solution with a known concentration is added (computer controlled) drop by drop to the digested solution. The potentiometric titration is based on the formation of precipitated silver chlorides, if chloride ions react with silver ions. The amount of silver nitrate solution needed to bind all chlorides can be determined by measuring the potential of an ion-selective electrode. Based on the content of silver ions, the total chloride content can be determined by stoichiometric calculations.

Table 3.4 Recommended number of investigation positions regarding corrosion of steel in concrete

Corrosion mechanism	Risk for reinforcement corrosion[e]	Recommended number of investigation positions (per 1000 m² concrete surface of areas with same conditions)[a]		
		Chloride concentrations[b]	Carbonation depths[c]	Inspection openings[d]
Chlorides	Critical area	≥5	—	≥3
	Uncritical area	≥3	—	≥1
Carbonation	Critical area	—	≥3	≥3
	Uncritical area	—	≥3	—

Source: European Federation of Corrosion Working Party 11, Corrosion of Steel in Concrete, *Materials and Corrosion* 64 (2013), no. 2.

[a] The values may be reduced if the summarised concrete surface area with the same material properties and exposure conditions is >10,000 m².
[b] Based on a full surface inspection regarding chloride-induced corrosion.
[c] Based on the combined evaluation of the concrete cover depth and the carbonation depth (e.g., phenolphthalein test).
[d] Determination of the loss of cross section and concrete cover depth.
[e] Risk for chloride or carbonation-induced reinforcement corrosion based on a full surface analysis of the structure.

Regarding the corrosion of steel reinforcement, the total amount of chlorides in the concrete is relevant, and not only the amount of chlorides that can be solved by water. The amount of chlorides that are water solvent is much lower than the total amount of chlorides, and so a dangerous underestimation can occur. It also has to be mentioned that all analytical methods lead to the chloride concentration relating to the mass of concrete. Usually the maximum chloride concentration is given in relation to the mass of cement. If the exact composition of the concrete is not known, the determined chloride concentration can be multiplied by a factor of about 8.

Regarding carbonation as well as the ingress of chlorides into the structure, the number of recommended investigation positions can be specified according to Raupach et al., 2013 (Table 3.4).

3.2.7.4 *Potential mapping*

Potential mapping is a feasible method to detect different levels of corrosion risks of steel in concrete. It is a nondestructive method that is suitable to detect areas with a high probability of corrosion induced by chlorides. It is not recommended for corrosion based on large depassivated areas with small potential gradients, as caused by carbonation. In contrast to corrosion induced by carbonation, chloride-induced corrosion is typically limited to local pits, as shown in Section 2.3.4. These corrosion pits lead to a negative potential of the reinforcement compared to the nondepassivated areas, which can be detected by potential mapping.

The so-called electrochemical potential is measured at the concrete surface by means of a high resistive voltmeter vs. a reference electrode (standard half-cell). Figure 3.54 illustrates the basic test setup. The positive terminal of the voltmeter has to be connected to the reinforcement, and thus locally the concrete cover has to be removed and any corrosion products have to be removed from the exposed reinforcement. Otherwise, the electrical connection between the voltmeter and the reinforcement can be affected negatively. The connection to the reinforcement can be done either with a gripper or by using a connection plug.

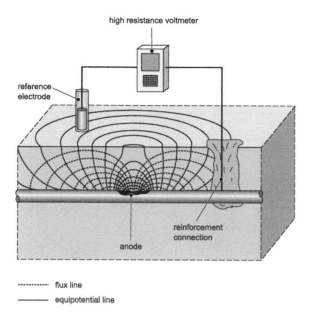

high resistance voltmeter

reference electrode

reinforcement connection

anode

---------- flux line
——————— equipotential line

Figure 3.54 Schematic drawing of the measuring principle of potential mapping.

The negative terminal of the voltmeter is connected to the reference electrode, which is then connected by an electrolyte to the concrete surface. According to ASTM C876, "the reference electrode selected shall provide a stable and reproducible potential for the measurement of the corrosion potential of reinforcing steel embedded in concrete." Several types of reference electrode are commercially available (ASTM C 876):

- Saturated copper/copper sulphate as the most commonly used electrode, which can be polluted by leaking of copper sulphate or by external pollutants, e.g., chlorides
- Saturated calomel (mercury/mercury chloride/potassium chloride) in saturated solution, which is comparatively fragile and not suitable for on-site investigations, but it can be used in laboratories
- Silver/silver chloride, which is usually used in marine environments

A diaphragm composed of wood or cork and a humid bridge using a wet sponge or felt usually realises the electrolytic connection between the reference electrode and the concrete surface. The sponge or felt has to be wetted regularly because otherwise the measurement is affected due to increasing electrical resistivity between the reference electrode and the concrete surface.

Depending on the type of measurement system used, more than one electrode can be measured at once. Usually wheel electrodes, shown in the Figure 3.55, are used to investigate large areas. Regarding on-site investigations, it is important that an uneven moisture distribution within the concrete affects the readings so that wet areas always have to be indicated in order to be able to interpret the results (Reichling et al. 2013). Depending on the type of electrode, a small water tank is integrated in the electrode, or as shown in the figure, the water is stored in an external tank, usually carried by the operator. It is also recommended

Figure 3.55 On-site potential mapping with a wheel electrode. Top: Single-wheel electrode with external water tank. Bottom: Four-wheel electrode with integrated water tank.

to carry out potential mappings at temperatures above 5°C, in order to avoid ice in the concrete or the reference electrode.

The measurements are usually done in a grid-like formation, such as the measurements of the concrete cover. Due to the funnel-like potential formed in the concrete (see Figure 3.54), large grid sizes will lead to a decreased probability of detecting active corrosion spots. Internationally, the recommendations of feasible grid sizes vary between 15 and 50 cm. Usually during on-site investigations the measurements start with a grid of 50 cm, and in critical areas the grid size is reduced to 10 cm.

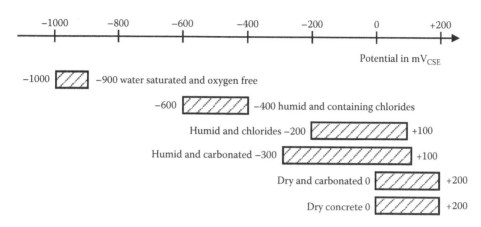

Figure 3.56 Typical potentials of reinforcing bars due to various environments. (From RILEM TC 154-EMC; Elsner, B., Andrade, C., Gulikers, J., Raupach, M., *Materials and Structures* (RILEM) 36 (2003), no. 261, pp. 461–471.)

As illustrated in Figure 3.56, the potential of the reinforcement decreases with an increasing chloride content and humidity of the concrete. Also, ongoing corrosion processes reduce the electrochemical potential of the reinforcement. So all mechanisms, except carbonation, that increase the risk of rebar corrosion of the reinforcement lead to a reduction of the electrochemical potential. This drop of the electrochemical potential can be detected by the potential mapping.

Extensive and systematic investigations carried out at the Institute of Building Materials Research (Bornstedt 1988) clearly indicate that no fixed potential value can be correlated with the risk of rebar corrosion. See Figure 3.57. For every single investigation the measured potentials have to be judged individually. Especially local gradients of the potential have to be regarded, and usually lead to areas with a high risk of corrosion or already ongoing corrosion.

Usually the results of a potential mapping are comprehended in so-called equipotential maps. These maps enable us to visualise areas with high and low potentials as well as steep gradients. In order to plan a full repair of a concrete structure in addition to potential mappings, the following investigations might be recommended:

- Visual inspection of the entire structure (see Section 3.2.2.2)
- Inspection holes in order to determine the status of corrosion (see Section 2.3.2)
- Determination of the concrete cover (see Section 3.2.3)
- Determination of the carbonation depth (see Section 3.2.7.2)
- Determination of chloride profiles (see Section 3.2.7.3)

The mentioned investigations, except the visual inspection, should be limited to areas in which the measured potentials show uncommon behaviour only. This enables an enormous reduction of the inspection costs because labour-intensive work as well as laboratory investigation can be reduced to a minimum without losing valuable information about the status of the structure, as shown in the following examples.

Figure 3.58 shows the outside (left) as well as the inside (right) view of a parking garage below a market square. The approximately 320 columns of a park deck below the market square were locally damaged by steel corrosion due to the ingress of chlorides.

A limited amount of columns were investigated and the investigations revealed no uniform results. Due to variations in the concrete quality, the exposure with chloride-contaminated

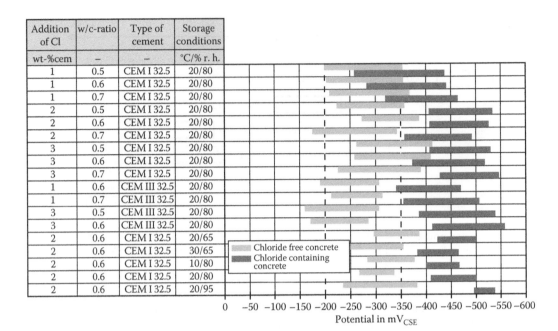

Addition of Cl	w/c-ratio	Type of cement	Storage conditions
wt-%cem	–	–	°C/% r. h.
1	0.5	CEM I 32.5	20/80
1	0.6	CEM I 32.5	20/80
1	0.7	CEM I 32.5	20/80
2	0.5	CEM I 32.5	20/80
2	0.6	CEM I 32.5	20/80
2	0.7	CEM I 32.5	20/80
3	0.5	CEM I 32.5	20/80
3	0.6	CEM I 32.5	20/80
3	0.7	CEM I 32.5	20/80
1	0.6	CEM III 32.5	20/80
1	0.7	CEM III 32.5	20/80
3	0.5	CEM III 32.5	20/80
3	0.6	CEM III 32.5	20/80
2	0.6	CEM I 32.5	20/65
2	0.6	CEM I 32.5	30/65
2	0.6	CEM I 32.5	10/80
2	0.6	CEM I 32.5	20/80
2	0.6	CEM I 32.5	20/95

Chloride free concrete
Chloride containing concrete

0   −50 −100 −150 −200 −250 −300 −350 −400 −450 −500 −550 −600
Potential in mV$_{CSE}$

Figure 3.57 Result of potential mapping under laboratory conditions. (From Bornstedt, H., Zerstörungsfreies Auffinden von korrodierender Bewehrung bei chloridbeaufschlagten Stahlbetonbauteilen mit Hilfe des Potentialmessverfahrens, Master's thesis, RWTH Aachen University, Institute of Building Materials Research, 1988.)

water (puddles vs. mist) as well as partially applied surface protection systems did not allow uniform results regarding the risk of corrosion due to chloride contamination. The initial plan suggested removing all concrete that might contain too many chlorides and applying a suitable repair mortar as well as a surface protection system. Due to the slenderness of the columns, the cross section would have been reduced up to 50%.

In order to reduce the costs of additional supports, it was intended to determine the chloride concentration of all columns and select the columns that need to be repaired. Due to various boundary conditions, the total amount of chloride analyses was greater than 1000. This number of analyses would have led to enormous costs as well as a time-consuming procedure. The realisation of a potential mapping combined with a limited amount of additional analyses turned out to be an economical and convincing solution.

The potential maps of the column footing were graphed as shown in Figure 3.59. On the left-hand side the northern side of the column is shown, and then counterclockwise the western, southern, and eastern sides are added. The potentials are mapped in 50 mV intervals beginning at −400 mV and ending at +100 mV. As already mentioned, the risk of corrosion increases with a decreasing potential.

In order to verify the potential measurements, selected columns were opened and the status of corrosion was judged. Comparing Figures 3.59 and 3.60, it can be clearly seen that the potentials as well as the state of corrosion correspond very well.

Complementary to these inspection holes the losses of cross section as well as the chloride profiles were determined and compared with the measured potentials. As expected, these investigations revealed that usually not all four sides of a column had to be repaired, but one or two sides. This enabled the repair of the entire columns without any additional supports.

Figure 3.58 Top: Outside view of a market square with a parking garage beneath. Bottom: Inside view of the parking garage.

Figure 3.61 shows the potential mapping of a trust-bearing wall of a bridge. The picture reveals low potentials at the edges that are due to chloride-contaminated water that ran down the walls.

In contrast to Figure 3.61, Figure 3.62 shows another trust-bearing wall with comparatively low potential values throughout the entire wall. These are also due to the rundown of chloride-contaminated water.

Also in multistory parking lots the chlorides are not distributed uniformly, but the chloride concentration within the concrete usually follows the use of the parking lot. Areas close to the entrance area, ramps, staircases, as well as highly frequented parking spaces, usually show a higher chloride concentration than areas with greater distances toward the entrance

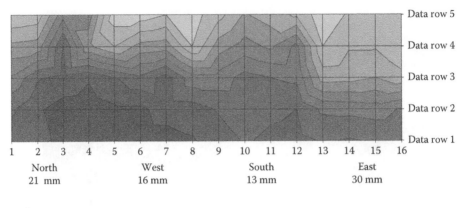

Explanation:

◻ −400–350 ◻ −350–300 ◻ −300–250 ◻ −250–200 ◻ −200–150

◻ −150–100 ◻ −100–50 ◻ −50–0 ◻ 0–50 ◻ 50–100 $[mV_{CSE}]$

Figure 3.59 Potential mapping of a column footing with comparatively low potentials in the northwestern corner.

Figure 3.60 Inspection hole at the northwestern corner revealing visible corrosion pits.

or the staircases. Figures 3.63 and 3.64 illustrate this. Parking deck C is very close to the entrance, and thus the lanes as well as the parking spaces show all very low potentials, and both are prone to chloride-induced corrosion. In the two stories above this deck (deck E), only the parking areas show low potentials, but not the driving lanes. This can be traced back to the fact that the cars already lost most of the chloride-contaminated water or snow in the lower decks, and only during the parking time is the water applied to the concrete.

Due to the individual use of every single building, these findings cannot be generalised. Figure 3.65 shows a parking lot where the driving lanes reveal a higher risk of corrosion than the parking areas.

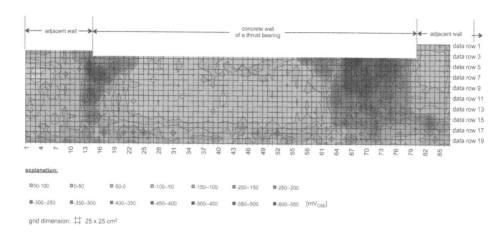

Figure 3.61 Potential mapping of a concrete wall revealing low potential values at the edges.

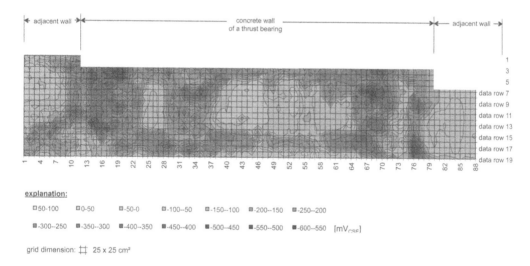

Figure 3.62 Potential mapping of a concrete wall revealing low potential values throughout the wall.

As shown before, conventional potential mapping needs a cable connection to the embedded reinforcement bars. Alternatively to the direct rebar connection, additional external electrodes may be used at the concrete surface to determine potential differences. To simplify the assessment the so-called Delta Sensor was developed (see Figure 3.66).

Between the three reference electrodes the potential differences are measured and a vector has been calculated that is displayed on a monitor or device. This vector points directly to the area with the most negative potential values, and the vector length correlates with the potential gradient. Thus, all information is given to locate adjacent areas with high corrosion risk.

Comparing the functionality with a compass, the user can move the sensor over the concrete surface, whereas the displayed vector permanently points to the critical spots relative to the current position. This procedure is advantageous in cases where, e.g., the concrete

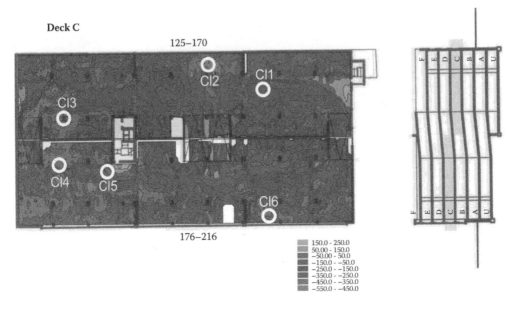

Figure 3.63 Potential mapping of a parking deck close to the entrance and nearly uniformly distributed low potential values.

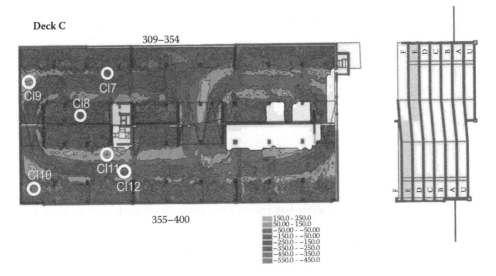

Figure 3.64 Potential mapping of a parking deck furthest to the entrance and low potential values in the parking areas only.

cover may not be removed locally or if several disconnected precast concrete elements are inspected. The method is currently used in research, and first measurements have been conducted at concrete structures successfully.

3.2.7.5 Electrical resistivity

The electrical resistivity is an important material property regarding concrete because of the following aspects:

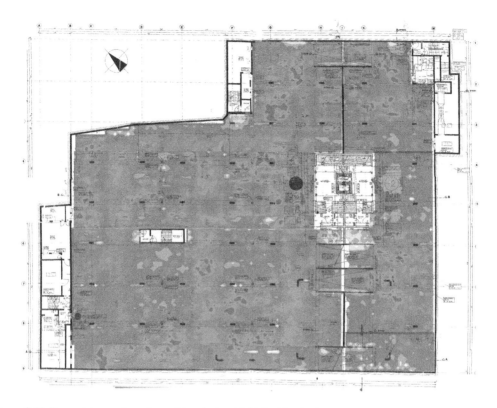

Figure 3.65 Potential mapping of a parking deck with high risk of rebar corrosion in the driving lanes.

1. The electrical resistivity of water-saturated concrete correlates with the resistivity against chloride ingress, so it can be used to judge the chloride diffusion coefficient.
2. If the reinforcement corrodes, the corrosion rate is inversely proportional to the electrical resistivity.
3. The electrical resistivity can be used to determine the water content of the concrete, if calibration data of the specific concrete are available.
4. During the design of cathodic protection systems the electrical resistivity of concrete is usually used to select adequate anode mortars as well as to simulate the current distribution within the structure (see Section 6.12.2).

The electrical resistivity of the concrete is based on the charge transfer capability of ions in the pore fluid.

Besides the material-based aspects, the readings are also influenced by the measurement setup itself. It can be distinguished using between two and four electrodes. In the first case, the same electrodes are used for injecting a current and for determining the resulting potential drop. Due to polarisation effects taking place at the interfacial surface between the electrodes and the concrete, the readings can be influenced. Therefore, the measurements should always be carried out using alternating current. Using a measurement frequency below a specific threshold value or using direct current results in an overestimation of the resistivity, which can lead to misinterpretations or useless readings.

The water saturation of the concrete can then be determined by using calibration curves as shown in Figure 3.67, which give a correlation between the resistivity and the saturation of a porous material, e.g., mortar or concrete. These calibration curves are determined in

Figure 3.66 Delta Sensor during operation on site. Top: Complete sensor. Bottom: View from the side.

the laboratory by using small samples with a defined water content and usually a two-electrode setup to measure the resistivity due to the exact saturation degree.

The correlation of the saturation as well as the recorded resistivity can be done by empirical law, e.g., by using Archie's law (Archie 1942; Reichling et al. 2012). The law is given by

$$\rho = \frac{\rho_F}{\Phi^m (S_w)^n}$$

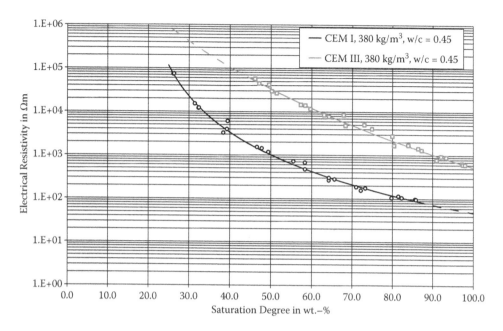

Figure 3.67 Correlation of the water saturation degree and the electrical resistivity for two exemplary concrete compositions.

where
S_W = Saturation degree
ρ_F = Resistivity of the pore fluid in the concrete in Ωm
ϕ = Porosity in vol%
m = Cementation index
ρ = Resistivity of the concrete for a certain saturation degree in Ωm
n = Saturation exponent

Archie's law allows gaining additional information about the pore structure of the material, and additionally the saturation degree exterior to the calibration range can be determined with certain reliability.

The given correlation between the water saturation degree and the resistivity is shown for two exemplary concrete compositions with different cement types. The regression analysis is based on Archie's law. It is obvious that both compositions show a different electrical behaviour based on the different pore structures.

Figure 3.68 shows the frequency dependency of the impedance for a concrete with different saturation degrees. It can be seen that the values depend strongly on the water content. The impedance values (alternating current resistivity) were measured with two electrodes. As a result of the polarisation effects at the electrodes, the values increase at low frequencies. High frequencies lead to an underestimation of the resistivity due to polarisation effects based on double-layer effects in the pore structure, the water content of the concrete, as well as coupling effects of the electrode cables. For this specific concrete the proper frequency range for the determination of the resistivity would be from approximately 10 to 1000 Hz. This range can differ for other concrete compositions. The influence of the polarisation can be omitted by using a four-electrode setup—see Wenner electrode.

In the laboratory the electrical resistivity can be determined by the two-electrode method, which is schematically shown in Figure 3.69. A cylindrical specimen with a known geometry

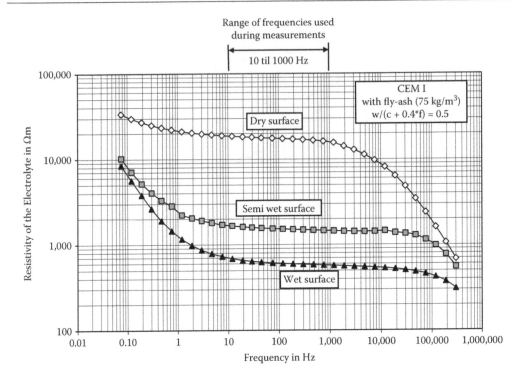

Figure 3.68 Electrical resistivity of a selected concrete in dependency of the water content as well as the measuring frequency. (Raupauch, Wolff, Dauberschmidt and Harnisch, 2007.)

is placed between two plane electrodes. Due to an applied voltage (alternating current), a uniform current distribution is achieved in the specimen (see Figures 3.69 and 3.70). The advantage of this test setup is that the voltage is applied in a uniform manner to the entire specimen, and so the influence of inhomogeneities is minimised. However, the coupling between the electrodes and the specimen should be done with extreme caution, and, e.g., dry electrodes can lead to measurement errors.

According to Ohm's law the resistivity of the specimen can be calculated based on the current and the voltage applied to the specimen. Usually not the absolute value of the resistivity, but the specific resistivity is calculated according to the following formula. The specific resistivity considers influences of the specific measurement setup, as well as the geometry of the specimen.

$$\rho = k \cdot R$$

where
ρ = Specific resistivity in Ωm
k = Cell constant in m
R = Absolute value of the electrical resistance of the testing cylinder in Ω

The so-called cell constant can be calculated by the following relation, which considers that the absolute value of the resistivity is proportional to the distance of the electrodes as well as inversely proportional to the cross section:

$$k = \frac{A}{d}$$

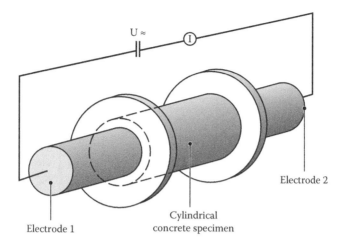

Figure 3.69 Schematic drawing of the two-electrode setup used to determine the resistivity of a building material.

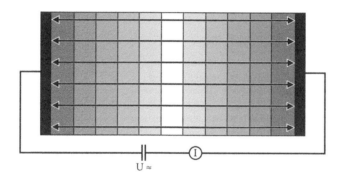

Figure 3.70 Uniform current distribution due to the two-electrode setup.

where

    k = Cell constant in m
    A = Perfused cross section in m^2
    d = Distance of the electrodes in m

Especially on-site measurements are usually done with the so-called Wenner method, which is done from the surface of the specimen. A detailed description of this test procedure can be found in, e.g., ASTM G 57-95a (2001) as well as Reichling et al. (2013). The Wenner test setup features four single electrodes (see Figure 3.71). The current feeding and the measuring electrodes are not identical. The outer electrodes induce a current into the material, and the inner electrode measures the potential drop in between.

The resistivity of the material can also be calculated with Ohm's law, but the cell constant has to be calculated with the following formula:

$$k = 2 \cdot \pi \cdot a$$

where

    k = Cell constant in m
    a = Distance of the electrodes in m

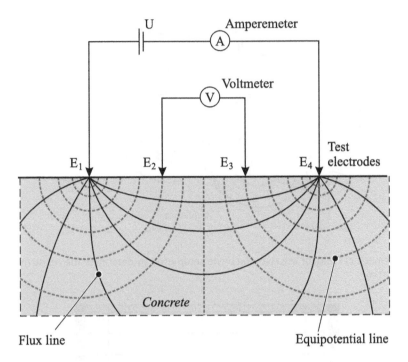

Figure 3.71 Schematic drawing of the Wenner test method.

Equal to the two-electrode test setup, the coupling of the electrodes also has to be done with caution. Usually the test devices using the four-electrode setup are equipped with felt-like tips in order to be able to wet the tips and achieve a good coupling between the electrodes and the concrete surface. A connected water film between the electrodes has to be avoided under all circumstances because otherwise a short circuit will be created.

Because the conductivity and hence the resistivity of steel and concrete differ in magnitudes, measurements above the rebars should be avoided. If the exact position of the rebars is unknown, it can be helpful to place the electrodes diagonal to the main reinforcement direction. Also, strong wetting of the surface and the material can alter the result of the measurement.

3.2.8 Water content and water uptake

3.2.8.1 General

The water content of parts of a building structure can deliver important information about the status of sealants as well as the water transport within the structure. Also, the water content gives valuable information about the risk of corrosion as well as possible ingress of chlorides due to water transport.

The water content can be measured directly on specimens taken out of the structure (drying method and CM method) as well as indirectly with nondestructive devices. All nondestructive methods have to be calibrated for every individual building material. Also, the water uptake can be determined on cores taken out of the building or nondestructively using Karsten's tubes.

3.2.8.2 Drying method

The drying method is a destructive investigation method that is comparatively simple to conduct and which delivers the absolute water content of a material. Even though the method is quite simple to conduct, it is usually not done on site because the measurement lasts between approximately three and five days.

In order to determine the absolute water content of a material, specimens have to be taken out of the structure. The amount of specimens varies with national standards as well as the desired result (see, e.g., (DIN) EN ISO 12570:2000). If the water content in dependency of the depth has to be known, e.g., if water from the rear of a wall has to be detected, the specimens have to be taken in different depths. During the extraction of the specimens, no additional water as well as heat should be induced into the material. Usually the specimens are not taken by drilling but with a caulking hammer, and directly after the extraction the specimens are wrapped in plastic foil.

After the extraction the weight of the samples is determined, and then the specimens are dried usually at 105°C until a constant mass is reached. If the material might contain polymers or gypsum, the drying temperature should not be above 70°C. This leads to an extended drying period, and not all water might be removed out of the material. Usually this mistake is acceptable because the driving out of polymers or changing the material due to a high temperature will lead to bigger mistakes.

The water content after the drying process can then be calculated by the following formula:

$$W_C = \frac{m_{wet} - m_{dry}}{m_{dry}}$$

where

W_C = water content
m_{wet} = Mass directly after extraction of the sample in g
m_{dry} = Mass after drying the specimen in g

It is recommended to use the drying method to calibrate all other—nondestructive—measuring techniques. After calibrating a nondestructive test method for an individual material these measuring methods can be used to enlarge the amount of measured values.

3.2.8.3 CM method

The CM method is a cheap and easy-to-conduct measuring technique that is commonly used for on-site measurements. The basic idea behind this measuring technique is that carbide and water react quickly to acetylene gas; see the following chemical equation:

$$CaC_2 + 2\,H_2O \rightarrow Ca(OH)_2 + C_2H_2$$

The amount of acetylene gas during the chemical reaction is equivalent to the water content of the specimen. It has to be noted that only the free water is measured during the test and not the chemically bounded water. The test itself is conducted in the following steps:

1. Crushing of the probe and weighting out a defined amount of crushed material. During the crushing process the humidity of the material should not be altered.
2. Inserting the crushed material and an ampoule, as well as steel balls, into a steel bottle, which is closed airtight. The bottle includes a pressure gauge.

3. Shaking the bottle for at least four intervals of 1 min every 5 min. After the shaking process the pressure within the bottle can be recorded and the humidity can be calculated by charts determined for different types of material.

The CM method features the big advantage that it takes only 20 min, and the result can be directly determined. The disadvantage of the CM method is that both the crushing and the shaking process can influence the results significantly, and thus lead to incorrect results.

3.2.8.4 Nondestructive devices

There are various nondestructive test methods available on the market that are based on measuring the water content indirectly by measuring, e.g., the conductivity or the capacity of the material. The conductivity, as well as the capacity, is usually measured in a two-electrode setup analogue to the Wenner method. The electrical current is induced into the material usually with two electrodes, and based on preinstalled calibration curves, the measuring gauge delivers a value of the water content based on the measured resistivity.

As long as the preinstalled calibration curves cannot be affected, the result should be judged with care because especially mineral-based materials, such as concrete, mortar, or stones, show great variations in the material properties, and so the achieved values might differ considerably from the correct value of the water content. Also, the measurements usually consider the border zone of a material, and thus any water gradients cannot be detected. The accuracy can be increased by determining calibration curves, e.g., with the drying method.

3.2.8.5 Water uptake

The water uptake of drill cores can be determined by immersing drill cores completely under water (so-called atmospheric water uptake—(DIN) EN 13755:2000) or by immersing the cores only with one surface up to 1 cm along the sides of the cores (so-called capillary water uptake—(DIN) EN 15148:2003).

In order to determine the water uptake, the weight of the cores has to be determined in regular intervals, e.g., 24 h, until the mass is constant. Then the cores are dried at 105°C until a constant mass is reached, and thus by the difference in mass, the water uptake can be determined.

If it is not possible to extract drill cores out of a structure, so-called Karsten's tubes can be used to determine the water uptake of a building material. Figure 3.72 shows a Karsten's tube, which is a glass tube with a height of 10 cm and a glass fitting. The glass fitting is connected to the surface by using dough. After connecting the Karsten's tube to the surface, the tube itself is filled with water and the reduction of the water level within the tube is recorded over time. Usually the test times are between 5 and 30 min. The water uptake of the building material is then given in $kg/m^2h^{0.5}$.

3.2.8.6 Sensors (MRE)

In order to measure the water content as well as the water distribution within the cross section of a structural element, sensors can be used that measure continuously the electrical resistivity of the material instead of the water content. The determination of the electrical resistivity can be done with more accuracy than the measurement of the water content. By using suitable calibration curves, the measured resistivity can be transferred into a water content of the material.

Figure 3.72 Top: Karsten's tube mounted on a brick wall. Bottom: Distribution of Karsten's tubes on a brick wall.

For more than 20 years the so-called multiring electrode (MRE) has been used to monitor the water content of the concrete as a function of the distance to the exposed concrete surface as a profile over time (see Raupach 1992). As most types of corrosion (frost, ASR, rebar corrosion, etc.) are to a considerable extent dependent on the water content of the concrete, long-term measurements (e.g., over one year) can give important information on the time dependence and maximum values of the corrosion rates.

The multiring electrode is made out of nine stainless steel rings, which are stacked together without any electrical contact in between. See Figure 3.73. Each ring—stainless

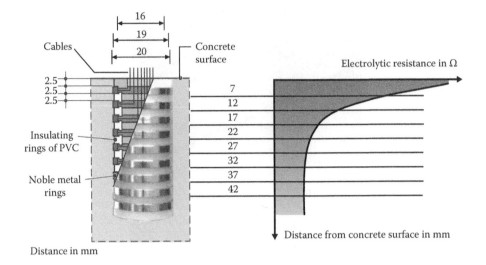

Figure 3.73 Left: Multiring electrode (MRE) as a monitoring sensor for the distribution of the water content of the concrete. Right: Schematic drawing of the resistivity distribution in over dependency of the depth.

steel as well as the plastic spacer—has a width of 2.5 mm. Each transverse axis of the rings is 5 mm apart, so that the resistivity of the material can be measured in depths between 7 and 42 mm. By applying an alternating current to each abreast ring, the resistivity can be determined between rings 1 and 2, 2 and 3, and so on until 8 and 9. The specific resistivity (see Section 3.2.7.5) can be calculated by using a cell constant of $k = 0.1$ m.

The installation of the MRE is done either by installing the sensors before placing the concrete or by placing the sensors into drill holes by using a special anchoring mortar. Both methods are displayed in Figure 3.74. In both cases the installation has to be done carefully so that the multiring electrode does not detach from the casting during the concreting, and the anchoring mortar has to be inserted without any air entrapment. The sensors are connected to a measuring and recording device.

The measurement of the resistivity (see Section 3.2.7.5) has to be done with an alternating current because the frequency does affect the measured values due to polarisation (see Catharin 1972; Warkus 2003).

The specific resistivity of concrete varies between roughly 50 Ωm for a water-saturated and hydrated concrete and above 10 MΩm for concrete that could dry off for a long time, e.g., interior building elements. Thus, concrete can be a fairly good electrical conductor but also an isolator. Figure 3.75 illustrates the time- as well as depth-dependent development of the resistivity of a dry concrete during the first three days of water storage. It can be noted that the water reaches the first measurement depth (7 mm) very quickly, while the inner layers need much more time until the water content rises, and thus the resistivity drops.

In order to interpret the results of the multiring measurements properly the following aspects have to be taken into account:

- The temperature has to be recorded and taken into account, e.g., by the Arrhenius correlation.
- The resistivity of concrete increases during the hydration process; thus, for young concrete the increase of the resistivity does not necessarily correlate with a drying process.
- The properties of the anchoring mortar have to be taken into account if the electrodes are placed into drill holes.

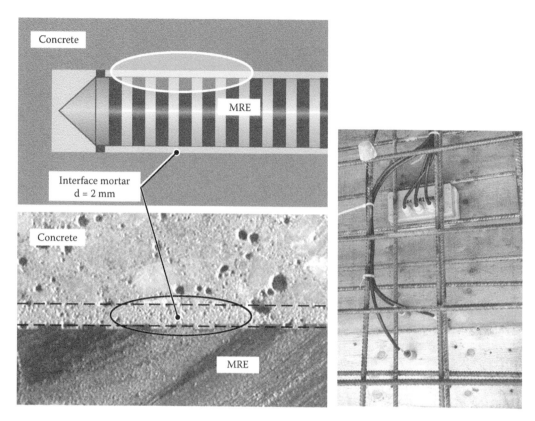

Figure 3.74 Schematic representation (top) of the installation of an MRE into the concrete by drilling a hole, inserting the MRE, and photographs of the gap filled with mortar (bottom) and MRE's installed before placement of the concrete (right).

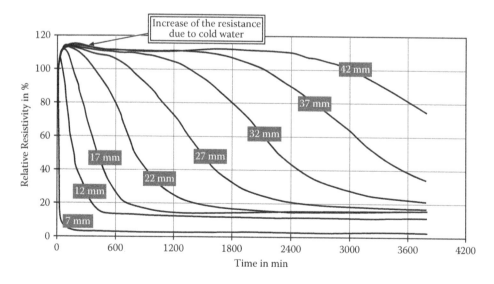

Figure 3.75 Time- and depth-dependent development of the resistivity due to water storage of concrete at 10°C after storage at 20°C/65% rh. (From Raupach, M., *Schriftenreihe des deutschen Ausschusses für Stahlbeton* (1992), no. 433, PhD thesis.)

Chapter 4

Evaluation of current and future conditions of the structure

4.1 GENERAL

The evaluation of the current condition of the structure and the prognosis of the further development is an essential task. It provides a decisive basis for the owner/operator and designer to select the optimal repair options for the structure. To select the technically and economically optimal time for possible interventions, a reliable estimation of the following points is necessary:

- Is the actual load-bearing capacity acceptable, and how will it develop in future?
- Is the actual usability and safety acceptable, and how will it develop in future?
- Is the concrete affected by corrosion, and how will the deterioration develop over time?
- Is the reinforcement subject to corrosion, and how will the deterioration (cracks, spalling, loss in cross section, etc.) develop over time?

For all points it has to be estimated when critical conditions will be reached, and critical limit states for the relevant processes have to be defined. These have to be discussed and agreed upon by the owner/operator and designer. Critical limit states can be defined in different ways:

- Critical conditions like crack width or deformation
- Limit value for safety factors
- Limit values for probabilities to reach critical situations

EN 1504-9 gives advice regarding assessment of defects and their causes:

An assessment shall be made of the defects in the concrete structure, their causes, and of the ability of the concrete structure to perform its function.
The process of assessment shall include but not be limited to the following:

a) the visible condition of the existing concrete structure
b) testing to determine the condition of the concrete and reinforcing steel
c) the original design approach
d) the environment, including exposure to contamination
e) the history of the concrete structure, including environmental exposure
f) the conditions of use (e.g., loading or other actions)
g) requirements for the future use.

The nature and causes of defects, including combinations of causes, shall be identified and recorded.

The approximate extent and likely rate of increase of defects shall then be assessed. An estimate shall be made when the member or concrete structure would no longer perform as intended, with no protection or repair measures (other than maintenance of existing systems) applied.

The results of the completed assessment shall be valid at the time that the protection or repair works are designed and carried out. If, as a result of passage of time or for any other reason, there are doubts about the validity of the assessment, a new assessment shall be made.

Regarding the factors to be considered to choose a management strategy, EN 1504-9 lists the following:

1) General
   a) the intended use and remaining service life of the structure;
   b) the required performance of the structure. This may include e.g., fire resistance and water tightness;
   c) the likely service life of the protection or repair works;
   d) the required availability of the structure, permissible interruption to its use and opportunities for additional protection, repair and monitoring work;
   e) the number and cost of repair cycles acceptable during the design life of the concrete structure;
   f) the comparative whole life cost of the alternative management strategies, including future inspection and maintenance or further repair cycles;
   g) properties and possible methods of preparation of the existing substrate;
   h) the appearance of the protected and repaired concrete structure.
2) Structural
   a) the actions and how they will be resisted, including during and after implementation of the strategy.
3) Health and safety
   a) the consequences of structural failure;
   b) health and safety requirements;
   c) the effect of occupiers or users of the concrete structure and on third parties.
4) Environmental
   a) the exposure environment of the structure and whether it can be changed locally;
   b) the need or opportunity to protect part or all of the concrete structure, from weather, pollution, salt spray etc., including protection of the substrate during the repair work.

In the following sections the possibilities to predict the future development of corrosion of concrete and reinforcement will be explained more in detail because there are no internationally agreed upon standard procedures like for the calculation of the load-bearing capacity. The series of standards EN 1504 also features no details on how to predict the progress of the different types of corrosion. General information and sophisticated approaches are given, e.g., in the actual fib-model code 2010 (fib 2010).

4.2 STATUS AND PROPAGATION OF CONCRETE CORROSION

The status of concrete corrosion can be determined using the methods described in Chapter 3. For an evaluation of the further development of the deterioration, corrosion models are required. These must take the relevant transport processes and reactions into account. However, the mechanisms of concrete corrosion are quite complex, and the accuracy of the existing models is limited.

In practice, as a first estimation it can be assumed that the deterioration progress continues at about the same rate as up to now. However, this might not be on the safe side because the starting time of the damaging process is often not clear, and the corrosion rates might accelerate with time, e.g., due to changing environmental conditions or the nature of the corrosion process.

For a more clear assessment, additional tests might be helpful. If sufficient time is available, the further development of the deterioration might be observed by regular visual inspections and measurements or using monitoring sensors. In cases where the extent of the damages is considerable and a prognosis of the further development is difficult, it is advisable to repair the damage immediately.

4.3 TIME TO REBAR CORROSION

4.3.1 General

For existing structures the time to corrosion of the reinforcement is always an important issue. To predict the corrosion rates, it has to be distinguished between corrosion induced by carbonation of the concrete and chloride-induced corrosion. For both cases the initiation and propagation phases have to be treated separately.

4.3.2 Carbonation-induced corrosion

In the case of carbonation-induced corrosion, usually only the initiation phase is considered: using models for the ingress of the carbonation front, it is calculated when the depth of the reinforcement is reached. The subsequent corrosion phase is usually not taken into account, but used to ensure safety against damages.

As a simple model for the ingress of the carbonation front, the following equation may be used:

$$X_c(t) = a^* \sqrt{t} \tag{4.1}$$

where
$\quad X_c(t)$ = Carbonation depth in mm
$\quad a \quad$ = Constant in mm/$\sqrt{a}$
$\quad t \quad$ = Time in a

This equation applies for concrete in a dry environment sheltered from rain. Figure 4.1 shows typical curves for this $\sqrt{t}$-law.

The constant a includes the diffusion coefficient of the concrete for CO_2. For usual concrete compositions it is in the range of 2–5 mm/$\sqrt{a}$, but for particularly good or bad quality (especially in old structures), it can be significantly lower or higher.

The $\sqrt{t}$-law may be used to estimate the time to depassivation by extrapolation of the depth of the carbonation front measured at a certain time to the future. As already mentioned, this $\sqrt{t}$-law is valid for dry environments. If the concrete surface gets wet, e.g., by rainfall, the diffusion rate of CO_2 is reduced, resulting in reduced carbonation rates. However, the time to corrosion will be calculated on the safe side.

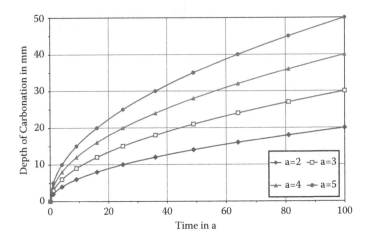

Figure 4.1 Carbonation depth over time for different constants *a* in mm/√a (√t-law).

To model carbonation under wet or periodically wet conditions, often the following law is used:

$$X_c(t) = a^* t^n \tag{4.2}$$

where

$X_c(t)$ = Carbonation depth in mm
a = Constant in mm/an
n = Constant, usually $0.25 < n < 0.5$
t = Time in a

For this law mathematically the square root of time ($n = 0.5$) is not used, but a root of higher order is more accurate, e.g., 3 or 4 ($n = 0.33$ or $n = 0.25$), which displays the ingress of the carbonation front under wet conditions. Besides this, the coefficient *a* will also be lower when the concrete is not continuously dry, because diffusion of CO_2 is reduced.

During recent years other, more sophisticated models have been developed, taking the properties of the concrete and environmental conditions into account in more detail; see, e.g., the fib model code (MC) 2010 (fib 2010). However, for the evaluation of concrete structures with unknown concrete mix designs special calibration tests are required. Generally experience from practice shows that the local distribution of the carbonation depth scatters quite distinctly, which makes accurate modelling difficult.

It can be expected that in the course of time the database will grow and commonly accepted test methods will be developed.

Usually the rate of carbonation-induced corrosion after depassivation is not taken into account. Anyhow, to predict the progress of deterioration, rates of the loss of thickness of the steel are used, which are known from experience with freely exposed steel. Mean rates of steel loss are in the range of 60 μm/a at 88% rh and up to about 300 μm/a at 100% rh. However, these values may be used as a rough estimate, but they are not directly transferrable to the conditions in carbonated concrete.

4.3.3 Chloride-induced corrosion

4.3.3.1 General

To predict the development of deterioration by chlorides, the initiation and propagation phases have to be considered. As already shown in the previous chapters, the relationships are much more complex than for carbonation-induced corrosion. Furthermore, for the definition of limit states the value of the critical corrosion initiating chloride content at the reinforcement is required. In the following sections the initiation phase, critical chloride content, and propagation phase are discussed.

4.3.3.2 Initiation phase

The initiation phase consists of the time of the ingress of chlorides until the critical chloride content is reached at the reinforcement.

To model the ingress of chlorides, commonly Fick's second law of diffusion is used. Theoretically this is not all correct because the prevailing transport mechanism in the pore system of the concrete is not only diffusion, but also capillary suction. Diffusion of chlorides occurs in water-saturated concrete, while in dry concrete water together with chlorides is sucked up by capillary action. However, experience has shown that chloride profiles can be described well using Fick's second law of diffusion. This means that the diffusion coefficient does not purely display diffusion, but also may include significant portions of capillary suction.

A full probabilistic design approach for the modelling of chloride-induced corrosion in uncracked concrete is presented in the fib MC 2010 (fib 2010). It is based on the limit state function, in which the critical chloride concentration C_{crit} is compared to the actual chloride concentration $C(x = a, t)$ at the depth of the reinforcing steel at time t. The change of the chloride content is given by the following equation:

$$C(x, t) = C_0 + (C_{S,\Delta x} - C_0) \left[1 - \mathrm{erf}\, \frac{x - \Delta x}{2 \cdot \sqrt{D_{app,C}(t) \cdot t}} \right] \qquad (4.3)$$

where

$C(x, t)$ = Content of chlorides in the concrete at a depth x (structure surface: $x = 0$) and at time t in wt%/cem

x = Depth in m

t = Concrete age in s

C_0 = Initial chloride content of the concrete in wt%/cem

$C_{S,\Delta x}$ = Chloride content at a depth of Δx in wt%/cem

Δx = Depth of the convection zone in m

$D_{app,C}$ = Apparent chloride diffusion coefficient in concrete in m²/s

erf = Error function

Often the concrete close to the surface is exposed to a frequent change of wetting and subsequent evaporation. This zone is usually referred to as the convection zone. In order to describe the penetration of chlorides for an intermittent load, Fick's second law of diffusion is applied starting with $C_{S,\Delta x}$ at the depth of the convection zone Δx. Experience has shown that with this simplification, Fick's second law of diffusion yields a good approximation of the chloride distribution at a depth $x \geq \Delta x$.

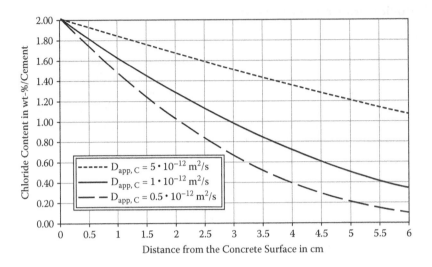

Figure 4.2 Calculated chloride profiles using Fick's second law of diffusion (30 a).

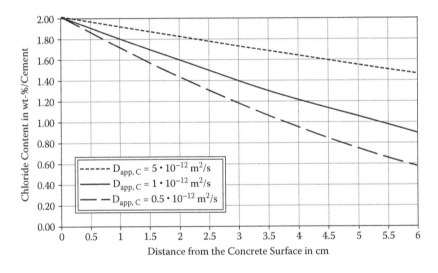

Figure 4.3 Calculated chloride profiles using Fick's second law of diffusion (100 a).

To demonstrate the influence of the diffusion coefficient, chloride profiles have been calculated using Fick's second law of diffusion for the assumption that the chloride concentration at the surface is 2 wt% related to the mass of cement, and that there is no convection zone ($\Delta x = 0$) (Raupach 2005). The apparent chloride diffusion coefficient has been assumed constant over time. Figures 4.2 and 4.3 show the results for different diffusion coefficients after 30 and 100 a.

Diffusion coefficients in the range of $5*10^{-12}$ m²/s are expected for concretes following the exposure class XD3 according to the European standards. Figures 4.2 and 4.3 show that a chloride content of 0.5 wt%/cem is already reached in a depth of 6 cm after less than 30 a for the given assumptions. The effect of the level of the diffusion coefficient can also be seen clearly.

To calculate the time to corrosion, Fick's second law of diffusion also can be used by plotting the depth of the critical chloride content over time. However, the level of the

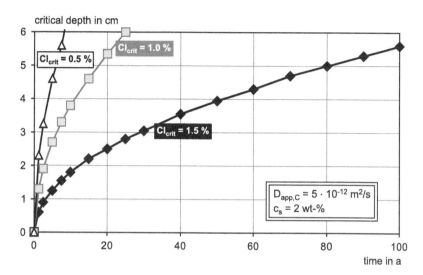

Figure 4.4 Critical depths calculated for a chloride diffusion coefficient of $5 \cdot 10^{-12}$ m²/s.

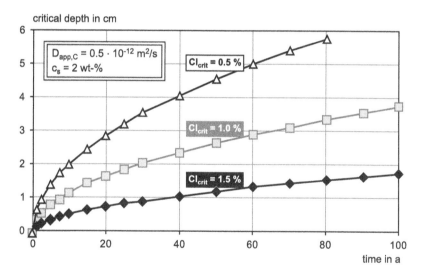

Figure 4.5 Critical depths calculated for a chloride diffusion coefficient of $0.5 \cdot 10^{-12}$ m²/s.

critical chloride content needs to be assumed. It is known that the corrosion initiating chloride content varies widely depending on several parameters. Therefore, three critical chloride contents C_{crit} between 0.5 and 1.5 wt% related to the cement content have been assumed for the calculations shown in Figures 4.4 to 4.5.

It is obvious from Figures 4.4 to 4.5 that not only the chloride diffusion coefficient $D_{app,C}$ and the surface chloride content C_s, but also the critical chloride content C_{crit} plays a decisive role for the prognosis of the time to corrosion. Therefore, in the following sections the level and influencing factors of the critical chloride content are discussed.

4.3.3.3 Critical chloride content

Experience has shown that the critical chloride content is not always a constant value as often assumed, but dependent on several parameters. To prevent corrosion problems in

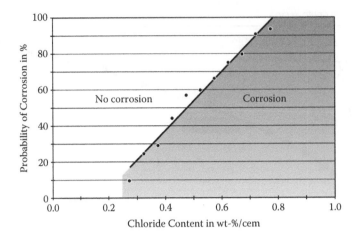

Figure 4.6 Statistical distribution of the critical chloride content regarding depassivation of the reinforcement according to Breit (1997).

European standards, the chloride content in the constituents of the concrete is limited to 0.4 wt%/cem. For safety reasons the maximum value for prestressing steels is 0.2 wt%/cem. It should be noted that these chloride contents are given as acid-soluble and not water-soluble concentrations.

For the assessment of existing structures different limit values are given in standards and regulations. Often a critical chloride content of 0.5 wt%/cem is assumed to be the limit value, where a detailed assessment regarding corrosion of the reinforcement is required. Some standards allow higher critical chloride contents for special concrete qualities and environmental conditions.

Often, it is not clearly stated for which limit state the critical chloride content is defined. It could be the limit value where depassivation occurs and corrosion starts. This is often used in research. Alternatively, the limit value is often used where first visible damages at buildings occur. This is usually relevant in practice. A review of the actual literature on the level of the critical chloride content is given in Angst (2009).

Breit (1997) has shown from extensive laboratory tests that the critical chloride content is not a constant, but statistically distributed with a quite high standard deviation, as shown in Figure 4.6. Below 0.25 wt%/cem no depassivation has been found. The mean value of the corrosion initiating chloride content was in the range of 0.5 wt%/cem.

In the fib MC 2010 (fib 2010) a similar distribution of the critical chloride content causing depassivation based on Figure 4.7 and further results is given by an asymmetric beta distribution with the following parameters:

- Mean value *m*: 0.6 wt%/cem
- Lower boundary *a*: 0.2 wt%/cem
- Upper boundary: 1.0 wt%/cem
- Deviation *s*: 0.1 wt%/cem

This distribution is shown in Figure 4.7 and is often used for service life design calculations as the limit state for the initiation phase. Besides these statistical considerations, research is still carried out to try to quantify the level of the critical chloride content for given conditions, allowing a more precise estimation of the corrosion risk, taking the relevant parameters into account.

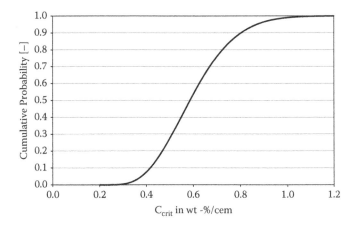

Figure 4.7 Distribution of the critical chloride content causing depassivation according to the fib model code 2010.

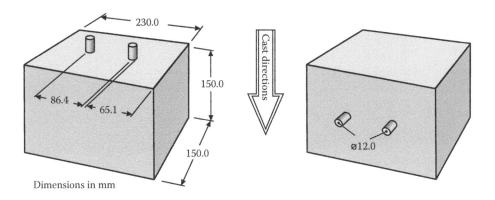

Figure 4.8 Specimen for laboratory tests on the influence of vertical or horizontal arrangement of the reinforcement on the critical chloride content.

It has been shown that besides the concrete mix design and environmental conditions, especially the amount of pores in the interfacial zone around the reinforcement influences the level of the critical chloride content (Glass and Reddy 2002). To quantify this effect, further research has been carried out, which is described in Harnisch and Raupach (2010). To simulate situations with a high and low amount of pores around the reinforcement, steel bars have been placed into the concrete standing (vertical) and lying (horizontal), as shown in Figure 4.8.

After hardening of the concrete, small specimens with a concrete cover of 10 mm on one side have been cut out of the concrete blocks. An artificial pore solution with a chloride content of 3 wt% has been applied to the side with the cover of 10 mm to simulate chloride attack. By means of electrochemical and chemical analyses, besides the direction of the rebars, also the impact of different concrete compositions on corrosion initiation times and chloride threshold levels was studied. Details are given in Harnisch and Raupach (2010).

The results are shown in Figures 4.9 and 4.10. A chemical analysis of the chloride content at the rebar level highlighted the fact that the influence of the rebar position dominates all other effects originating from different mix properties, such as the water/cement (w/c) ratio

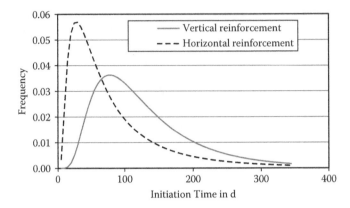

Figure 4.9 Statistical evaluation of the initiation times of all specimens with regard to the rebar position.

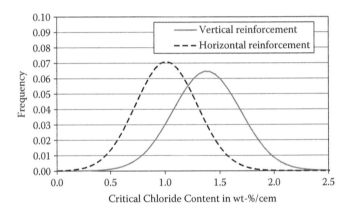

Figure 4.10 Statistical evaluation of the critical chloride contents of all specimens with regard to the rebar position.

or the filler content. Thus, consistently higher chloride threshold levels could be detected for probes with a vertical arrangement of the reinforcement.

The results clearly demonstrate that obviously the interfacial zone between the reinforcement and the surrounding concrete has a pronounced influence on the critical chloride content, and consequently also on the initiation times. Furthermore, they show that the critical chloride content is not limited to 1 wt%/cem, as indicated in Figure 4.7, but can be significantly higher. Further research on this topic is still ongoing.

4.3.3.4 Propagation phase

Often the propagation phase of chloride-induced corrosion is not taken into account for service life designs because as already shown, the corrosion rates are usually much higher than for carbonation-induced corrosion. Therefore, in many cases the propagation phase is quite short compared to the initiation phase. It has been shown that critical limit states can be reached within only a few years.

On the other hand, in aggressive environments containing chloride concretes with high quality and high covers, today so-called high-performance concretes are used. It can be expected that the propagation phase for such concretes is not negligible and worth consideration for the prognosis of the development of deterioration. However, special care has to be taken in the area of cracks in concrete, where macrocell corrosion occurs.

Using modern optical scanning techniques with high resolution, the development of corrosion on the surface of steel has been investigated (see, e.g., Harnisch and Raupach 2011). Macrocell specimens with small anode bars (diameter 10 mm, length 50 mm) in concrete containing 3 wt%/cem and cathode bars with a 50 times larger surface in chloride-free concrete have been used. The concrete surface at the anode bar was set under water for one month, followed by another month without water, etc. This simulates quite harsh corrosion conditions. However, in extreme cases the situation in practice may be even worse.

After scanning of the probes by means of optical triangulation, a digital 3D image of the corroded steel surface is available. In order to simplify the evaluations, the corroding surface areas have been unrolled by coordinate transformation into a plane, as shown in Figure 4.11.

From these digital data the amount of corroded area in every depth can now easily be quantified, and an in-depth profile of the corroded area can be extracted, as shown exemplarily in Figure 4.12. For this analysis the reference plane was moved in steps of 5 μm until no more corroded areas could be detected.

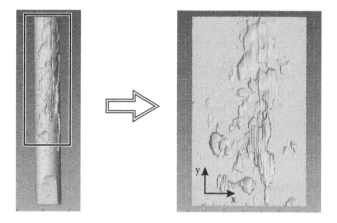

Figure 4.11 Transformation of selected bar surface areas for analysis.

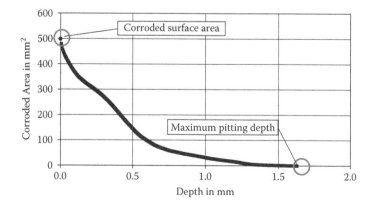

Figure 4.12 Example for an in-depth profile of the corroded area of a steel bar.

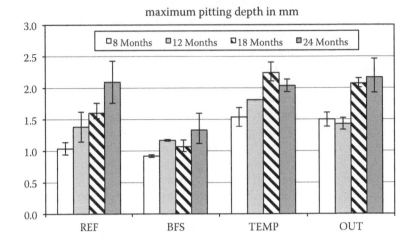

Figure 4.13 Maximum pitting depths of steel in different concretes over time.

It can be seen that as expected, the corroded surface area decreases over depth until the maximum pit depth is reached. These data are an interesting basis for modelling of the mass loss of the steel over time.

Figure 4.13 shows the maximum pitting depths measured at different specimens as described in Harnisch and Raupach (2011). The series called REF consisted of concretes with CEM I 42.5, w/c ratio 0.5, and a cover depth of 20 mm. For the series BFS blast furnace slag (CEM III/B) was used instead of OPC. For the series TEMP the temperature was 30°C instead of 20°C, and the series OUT was stored outdoors in Aachen, Germany.

Figure 4.13 shows that the maximum pitting depth was already about 1 to 1.5 mm after eight months and increased to more than 2 mm after two years, except for the concrete with BFS, which is known to have a high electrical resistivity.

Regarding the load-bearing capacity of the reinforcement, not only the pit depth is relevant, but primarily the loss of cross section, which can also be extracted from the 3D data. While the remaining cross section is relevant for ductile materials like reinforcing steel, the pit depth is more relevant for the load-bearing capacity of brittle materials. Figure 4.14

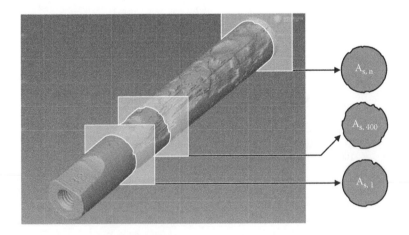

Figure 4.14 Determination of the remaining cross sections of the corroded steel bar.

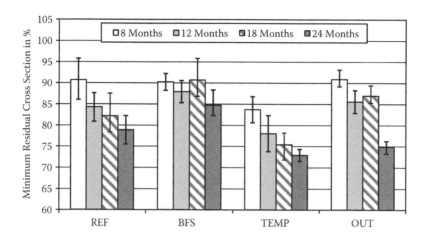

Figure 4.15 Minimum residual cross sections in different concretes over time.

shows schematically how the remaining cross sections after corrosion are determined along the steel. They vary considerably along the steel axis.

For the bars with 10 mm diameter the initial cross section without corrosion is 78.5 mm². Figure 4.15 shows how this cross section is reduced over time. It is shown that after eight months the minimum residual cross section is about 90%, except for the TEMP series where it already was reduced to less than 85%. After two years the minimum residual cross sections were further reduced, but differently for the investigated series. For the series TEMP and OUT the residual cross sections were about 75%, or even less already after two years. For the series with BFS, as expected the loss of cross section was least of all.

The results show that under aggressive conditions the residual cross section may be reduced to 75% at a propagation time of only two years in noncracked concrete. However, it has to be considered that after depassivation, in practice the chloride content is not in the range of 3 wt%/cem, but lower. So these results are not directly comparable with the situation in practice, but they give an idea of how quick the cross section can be lost due to chloride-induced corrosion.

The acceptable reduction rate of the cross section of the concrete depends on the actual load factor of the structural element and has to be attested.

In the tests described above, a nominal cathode/anode surface area ratio of 50 has been used to simulate conditions with high corrosion rates. This raises the question of how the geometrical arrangement of reinforcement acting anodically and cathodically influences the corrosion rates.

To investigate this question, extensive laboratory tests and numerical simulations have been carried out (Warkus 2012). It has been shown that macrocell action today can be simulated numerically by special software tools quite exactly. Figure 4.16 shows selected results of a numerical parameter study on the influence of the geometry of corrosion systems consisting of depassivated anodic areas and passive cathodic areas. The assumptions for the driving voltage U, the anodic polarisation resistance $r_{P,A}$, and the electrical resistivity ρ given in Figure 4.16 simulate again parameters for quite high corrosion rates (Warkus 2012).

An electrical current of 1 µA/cm² corresponds to a mean loss of thickness of the steel of 11.6 µm/a. The numerical simulation has been carried out for different cathode/anode area ratios as indicated in Figure 4.16. The results show that the highest macrocell corrosion rates are expected for a local depassivated area of a slab with a cathode/anode area ratio of 500 with a mean rate of 7.2 µA/cm² equivalent to 84 µm/a.

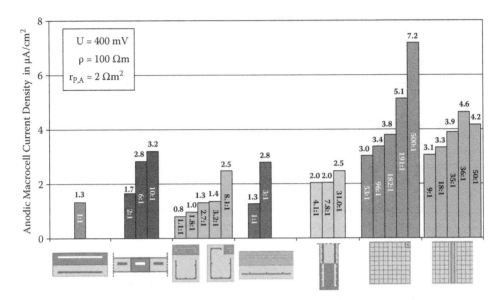

Figure 4.16 Macrocell corrosion rates for different geometrical arrangements of anodes (dark) and cathodes (light).

It has been shown that the smallest remaining cross section is about a factor of two or three smaller than the mean value of the remaining cross section (Harnisch and Raupach 2011). Furthermore, besides macrocell corrosion, also the effects of microcells (self-corrosion) have to be considered. This means that a mean rate of 84 μm/a according to Figure 4.16 corresponds to a considerably higher maximum rate of steel removal.

Calculations have shown that for the condition of corrosion in cracks in concrete the maximum rates of steel removal are in the range of 80 μA/cm². This corresponds to about 1 mm/a (Warkus 2012). An example for such high corrosion rates in cracked concrete attacked by chlorides has already been shown in Section 2.3.4.

Figure 4.16 shows significantly different corrosion rates for different types of slabs, beams, or a column, typical arrangements of depassivated areas (anodes) and passive reinforcement (cathodes). This means that the geometrical conditions of a structure need to be considered for the estimation of the propagation phase, and that the propagation phase is not necessarily very short.

Initiation and propagation phases can be treated together in combined probabilistic models for service life design and also for the prognosis of the condition of existing structures (see, e.g., Kosalla and Raupach 2012).

Figure 4.17 shows exemplarily for the parameters given in Kosalla and Raupach (2012) reliability indices for the initiation phase alone with $C_{Crit} = 1.0$ wt%/cem, the propagation phase alone with critical loss in cross section $\Delta A_s = 19\%$, and for the combined deterioration model.

It shows as expected that the reliability increases when the corrosion phase is also taken into account for the deterioration model. However, the acceptable minimum value for the reliability also has to be increased when taking the corrosion phase into account. The acceptable minimum reliability has to be defined with great care because it includes the safety factor of the structure for durability. As the knowledge regarding corrosion behaviour of concrete structures is increasing by research and practical experience, it can be assumed that reliability calculations will become more and more attractive in the future.

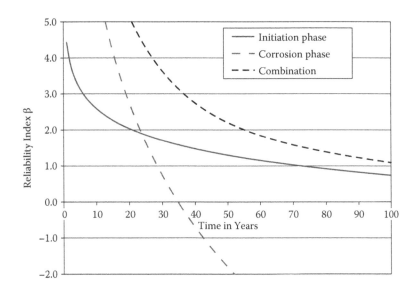

Figure 4.17 Development of the reliability index β for the initiation and propagation phase and the combined deterioration model. (From Kosalla, M., Raupach, M., in *Safer World through Better Corrosion Control, EUROCORR 2012*, Istanbul, September 9–13, 2012, Istanbul: EFC, 2012.)

Chapter 5

General procedures and materials for repair

5.1 GENERAL

This chapter describes the general procedures and types of materials for repair and protection of concrete structures. For most methods, as one of the first steps of the execution of the repair works, the concrete surface has to be prepared.

5.2 PREPARATION OF THE CONCRETE SURFACE

The preparation of the concrete surface is one of the most important steps in making a successful concrete repair. The primary goal of the surface preparation is a good and durable bond between the repair material and the substrate. As a rule of thumb, the substrate has to be sound, clean, rough textured, and dry after the surface preparation. Figure 5.1 demonstrates this rule of thumb; the adhesion between the concrete—which is part of a bridge column—and the repair mortar (40 mm of SPCC) was not sufficiently high, so that large areas of the repair material delaminated and fell off the concrete. This delamination not only compromises the repair works, but also can harm people, who are located below the bridge, as well as the columns. The delamination occurred even though the strength of the concrete, as well as the SPCC, was very high.

Depending on the boundary conditions, the following steps are done to prepare the surface:

- Dry cleaning, e.g., sweeping
- Removal of thin layers with low strength, e.g., cement slurry
- Removal of concrete or mortar
- Removal of old coatings or impregnations
- Removal of rust located on the reinforcement

The surface preparation has to be done in order to achieve a substrate, which fulfils the following aspects:

- No parts of low strength or insufficient adhesion, e.g., flanks of cracks with low strength, as well as no parts of cement slurry. Furthermore, the surface should not sand off or chalk.
- No peel-off or crack running parallel to the surface, as well as cracks or separations close to the surface.
- Free of grouts. Grout might only be allowed in justified cases on behalf of the planer.
- A sufficient roughness that is according to the requirements of the repair material.

Figure 5.1 Large delamination of repair mortar on a bridge column.

- Free of any impurities, e.g., rubber dust, oil, gasoline, wax, releasing agent, old coatings or impregnations, moss, or other vegetation.
- Voids or defects that are not repaired properly.

The specific requirements, e.g., roughness, adhesion strength, or moisture content, are usually given in the specific data sheet of the repair material that will be used for the concrete repair.

In order to achieve a sufficient surface preparation, various methods can be used. Table 5.1 gives an overview of selected and most commonly used methods according to RL-SIB (2001).

Figures 5.2 to 5.4 show various methods for cleaning and preparation of the concrete surface. Cleaning can also be done by using dry ice, as shown in Figure 5.2. The advantage of cleaning with dry ice is that no water is forced into the concrete. But due to the emerging of CO_2, personal safety equipment in combination with proper ventilation (indoor use only) is mandatory. Regardless of what type of cleaning is executed, the surface is only cleaned and no material is taken off the surface.

Unlike cleaning, Figure 5.3 illustrates the difference effect on the reinforcement of manual concrete removal with a caulk hammer and high-pressurised water jetting (Figure 5.4), respectively.

Figure 5.5 shows a bridge deck after concrete removal by milling as well as a milling machine. It has to be noted that the milling process itself can damage the reinforcement, which is too close to the surface, severely, as also shown in Figure 5.6. This disadvantage can be avoided if the concrete is removed by high-pressurised water jetting. After the milling process, the concrete surface has to be treated by shot blasting in all cases. The difference between a milled and a milled and shot-blasted surface is shown in Figure 5.8.

The shot-blasting process as well as the concrete before the required cleaning is shown in Figure 5.7. The shot blasting is done with small steel balls that partly remain on the surface of the concrete even though the balls are sucked off the surface by a magnet in the shot-blast machine. The remaining steel balls are then removed by a magnetic roll or, if other materials are used, by a vacuum cleaner.

As mentioned before, after milling a concrete surface the surface has to be further prepared by shot blasting. The difference between the two steps of surface preparation can be clearly seen in Figure 5.8. It can be seen that the surface is more even due to the shot blasting, and the milling grouts are removed entirely.

Table 5.1 Overview of selected and most commonly used methods for surface preparation according to RL-SiB (2001)

Procedure			Used for[a]					Typical application	Requirement	Required postprocessing
Type	Tools		1	2	3	4	5			
1	2		1	2	3	4	5	4	5	6
Mechanical impact	(Caulking) hammer	By hand	x	x	x			Small areas[b]	Damage of the reinforcement; especially prestressing steel has to be avoided	Shot blasting
	Chisel	Electric or pneumatic								
	Needle scaler		x	x		(x)				
Brush off	Rotating steel brush		x	x		(x)		Depending on type of tool		Cleaning
Milling	Milling machine		x	x	x			Large removal of ≤5 mm in each step; large areas require automatic levels	Usually removal of ≤5 mm in each step; large areas require automatic levels	Shot blasting
Grinding	Grinding machine		x	x				Small areas		Cleaning
Flame cleaning	Flame blasting equipment[c]		x	x				Vertical and horizontal surfaces		Cleaning
Dustless shot blasting	Shot blasting with additional vacuum cleaner		x	x	(x)	x		Vertical and horizontal surfaces—depending on the equipment used		
Shot blasting	Shot blasting with pressurised air		x	x	(x)	x		Vertical and horizontal surfaces	Dust protection; oil-free pressurised air	Cleaning
	Shot blasting with moisture-blasting abrasive		x	x	(x)	(x)[e]			No dust protection required; oil-free pressurised air	
	High-pressure water jetting		x	x	(x)[d]	(x)[e]			No dust protection required; oil-free pressurised air	
Cleaning	Pressurised air		x	x			x	No horizontal surfaces[b]		
	Vacuum cleaner		(x)				x	Horizontal surfaces		
	Water jetting						x	Removal of, e.g., moss or similar impurities		

[a] Used for: 1 = Removal of impregnations, coatings, or curing agents; 2 = Removal of cement slurry; 3 = Removal of weak concrete and exposure of the rebars; 4 = Removal of rust; 5 = Cleaning of the concrete surface.
[b] Risk of damaging sound concrete.
[c] Thermally destroyed concrete has to be removed.
[d] Coatings might not be removed completely.
[e] Might require shot blasting as a second step.

Figure 5.2 Cleaning of a concrete surface by using dry ice.

Figure 5.3 Reinforcement after concrete removal. Left: Concrete removal with a caulking hammer. Right: Concrete removal by high-pressurised water jetting.

The preparation of the surface has to be done until the required surface strength is reached in order to ensure a proper and durable adhesion between the repair material and the substrate. Otherwise, as shown in Figure 5.1, a durable adhesion is not granted even though both materials reach the designated material properties. Usually the surface strength of concrete before applying mortar or polymeric surface protection systems has to be at least

Figure 5.4 Concrete removal by high-pressurised water jetting on a bridge deck.

1.0 N/mm² (singe value) or 1.5 N/mm² (mean value of all test spots). Figure 5.9 illustrates the dependency between the compressive strength and the surface strength of concrete. It can be seen that especially concrete with a compressive strength lower than 20 N/mm² does not necessarily have a surface strength greater than the mentioned values.

5.3 REPLACEMENT OF DAMAGED CONCRETE

5.3.1 Introduction

Depending on the condition of the structural concrete or the contamination of the concrete with chlorides, as well as the depth of carbonation, usually a certain amount of concrete is damaged due to corrosion of the concrete or rebars, or it was removed during a repair. These breakouts have to be replaced with a suitable material. Generally speaking, the following materials can be used to repair concrete:

- Cement mortars or concrete (CC)
- Sprayable cement mortar or concrete (SPCC)
- Polymer-modified mortars or concrete (PCC)
- Polymer mortars (PC)

The following paragraphs give a short overview of the differences between the materials.

5.3.2 Concrete and sprayed concrete

If the thickness of the required concrete repair is large, usually concrete (according to (DIN) EN 206:2012) as well as sprayed concrete (according to (DIN) EN 14487:2006) is used to repair concrete structures. Concrete as well as sprayed concrete features a minimum aggregate size of 8 mm and is usually applied in layers with a thickness of at least 30 mm (sprayed concrete) or 50 mm (cast concrete).

Figure 5.5 Concrete removal with a milling machine. Top: Machine during operation. Bottom: Detail of the milling head.

Besides the thickness of the repair patch, the orientation also contributes to the material selection. *In situ* placed (or cast) concrete is only used for the repair of horizontal surfaces (usually without using a formwork) or vertical surfaces (with formwork). *In situ* placed concrete cannot be used for any overhead surfaces. In contrast, sprayed concrete can be used for all overhead and vertical surfaces, but never for horizontal surfaces.

If cast concrete can be used, all requirements regarding the composition of the concrete due to the exposure as well as the load-bearing behaviour have to be considered during the design process. In Europe these requirements are all comprehended in (DIN) EN 206:2012 (composition of concrete) as well as Eurocode 2 (structural design). For patch repair, concrete might also be placed manually, as shown in Figure 5.10.

One crucial aspect of placing concrete in front on an existing structure is the adhesion between the old and the new concrete. Usually a sufficient adhesion is realised with creating a rough surface by shot blasting, abrasive blasting, or hydroblasting. Often additional anchors between the old and the new concrete are required. Using a formwork (repair of

Figure 5.6 Concrete removal with a milling machine. Top: Concrete surface after milling process. Bottom: Detail of the damaged reinforcement.

vertical surfaces) enables the designer to specify a particular surface analogue to new buildings. Depending on the orientation of the casting, the concrete is either poured or pumped into the formwork. Small defects are usually hand-filled. In all cases a sufficient compacting—equivalent to placing new concrete—is required.

In contrast to cast concrete, sprayed concrete does not require an extensive formwork but is applied by spraying a premix onto the surface. Also, due to the spraying process, no additional compaction is required. The composition as well as the design of sprayed concrete has to be done according to (DIN) EN 206:2012 and (DIN) EN 14487-2:2006, which

Figure 5.7 Surface prepration by shot blasting.

regulate certain additional material properties, e.g., water uptake and adhesion strength on the surface. Additionally, sprayed concrete can be reinforced with fibres in order to increase the bending strength and reduce the risk of cracks due to shrinkage. After placing the concrete, the concrete has to be cured just like regular concrete surfaces used for building new constructions.

5.3.3 Cement mortars

For cement mortars the maximum grain size is limited to 4 mm and the applied layers have a thickness between 20 and 40 mm. Cement mortars are regulated in the EN 196 series and also have to meet all the requirements due to exposure as well as load-bearing behaviour. Cement mortars are usually used to create thinner layers in comparison to concrete layers. The application can be done either by placing in formwork (see Section 5.3.2) or by spraying. If cement mortars are applied by spraying, polymers are added to the dry mix, and these materials are called SPCC (see next section).

5.3.4 Polymer-modified mortars PCC and SPCC

Polymer-modified mortars are based on cement mortars that additionally contain polymers. These mortars can also be applied by hand (polymer cement concrete (PCC); see Figure 5.12) or by spraying (sprayable cement concrete (SPCC; see Figure 5.11)). The selection of the application method has to be done considering the orientation of the surface.

Horizontal surfaces or surfaces with only a slight slope can only be repaired with a PCC. SPCCs cannot be used there because due to the spraying, a certain amount of mortar (mostly aggregates) bounces off the surface (so-called rebound). This rebound has to be considered as waste and cannot be included in the mortar, which will automatically happen on horizontal surfaces. On vertical surfaces or on ceilings the application of SPCC is a feasible technique because the rebound always falls down and is not included in the mortar.

Figure 5.8 Top: Detailed view of a concrete surface after milling. Bottom: Detailed view of a concrete surface after milling and additional shot blasting.

Usually the polymer content in the dry mix of (S)PCCs is between 0.5 and 5%, and limited to a maximum of 10% if the mortar is used for realkalisation, because only the cement and not the polymer contributes to the alkalinity (see Method 7.4 in Section 6.9.5) of the (S)PCC.

The main reasons for adding polymers in comparison to pure cement mortars are:

- Increase of the adhesion to the surface
- Advanced workability
- Increase of water retention
- Increase of bending and tension strength
- Reduction of Young's modulus

The increased adhesion to the surface as well as the advanced workability is mainly relevant during the application of the material. The increased water retention reduces the

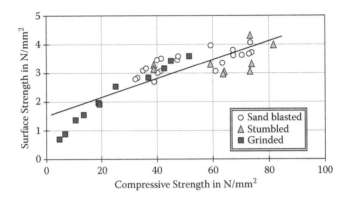

Figure 5.9 Surface strength vs. compressive strength for different types of concrete as well as surface preparations. (From Schulz, R.-R., *Beton als Beschichtungsuntergrund: Über die Prüfung des Festigkeitsverhaltens von Betonoberflächen mit dem Abreißversuch*, Dissertation, Technische Hochschule, Aachen, Fachbereich 3, 1984.)

sensibility to environmental impacts during application as well as curing. The reduced mechanical strength as well as the reduced Young's modulus reduces the risk of shrinkage cracks and detachments due to deformation of the existing structure.

The polymers added to the mortar mix are usually thermoplastic polymers in the form of dispersions, e.g., acrylics, styrene-acrylic copolymers, styrene-butadiene polymers, or vinyl-acetate polymers. But also flow agents or defoaming agents are added to the PCC or SPCC mixtures. The polymers are usually added to the dry mix by the manufacturer. During the hydration phase the dispersions dry and form polymeric networks that are distributed homogeneously in the pore structure of the mortar.

For PCCs the amount of water to be added is usually specified in the technical documentation. The mortar itself is mixed in a mixer capable of mixing cement-based materials with a grain size below 8 mm.

If applied by spraying, two different application techniques have been established. SPCCs can be applied by dry or wet spraying. The mortars are then called analogue to the application technique—dry-mix shotcrete or wet-mix shotcrete. Mortars applied by wet spraying are also premixed, and the wet mortar mixture is pumped to the spray gun by a shotcrete machine. Dry-sprayed mortars are pumped to the nozzle without water, and the water is directly added in the nozzle by the operator. This enables the operator to adjust the water content to the boundary conditions, such as temperature, humidity, or the amount of material bouncing of the surface. This freedom of the operator's decisions during the application requires trained personal as well as the control of the material properties during the application. The training of the application personal is crucial because the operator on the nozzle finally adjusts the water/cement ratio, which is decisive to reach the desired properties. Usually during the application the water content of the mortar is determined by the drying method (see Section 3.2.8.2) and compared with the requirements of the manufacturer in order to avoid an incorrect composition of the mortar.

The surfaces of PCC repair patches are usually trowelled by hand and thus smoothed. SPCC repair patches usually remain rough as applied. If a smooth surface is required, SPCC usually has to be applied in two layers. The second layer can be made out of SPCC or a specific fine speckle (see Section 5.3.6). If the second layer is made out of SPCC, on top of

Figure 5.10 Top: Manual application of concrete realising a patch repair on a bridge deck. Bottom: Finished patch repair; partly still curing by linen applied.

the first layer—applied as required by the manufacturer—the second layer is applied with a higher water content that enables the smoothing of the surface. This layer should only be considered as an additional layer that ensures the optical quality and does not contribute to the required layer thickness regarding protection and durability.

5.3.5 Polymer mortar PC

In contrast to concrete, cement mortars, or polymer-modified mortars, polymer mortars (PC) do not contain cement as a binder but only polymers. Usually PCs contain reactive

Figure 5.11 Application of SPCC (dry-mix shotcrete) test samples.

Figure 5.12 Application of a PCC by hand.

polymers such as epoxy resins. The reactive polymers are always cold hardening and alkali stable. PCs should only be used in exceptional cases, e.g.:

- Fast-drying patches required
- No curing treatment possible
- Extremely low thickness required

PCs are usually delivered in two-component containers with defined ratios because without the correct so-called stoichiometric ratio, reactive resins do not react in the desired manner. Thus, the intended material properties are not reached. Also, the aggregates are usually delivered by the manufacturer of the PC.

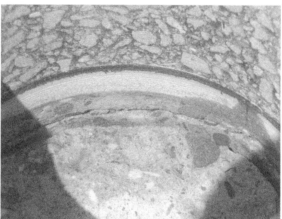

Figure 5.13 Delamination of a PC due to a large thickness. Top: Delamination of the PC. Bottom: Damage of the concrete below the PC layer.

PCs are only applied in thin layers, which feature greater adhesion strengths to the substrate than polymer-modified concrete. If PCs are applied in thick layers or the substrate does not have a sufficient strength, the difference in the thermal expansion coefficient may lead to severe damages (see Figure 5.13). The examples show that not only the PC delaminates from the concrete surface, but also the concrete might be damaged. To prevent such changes, the size of patches filled with PC should be limited to <1 m^2.

5.3.6 Primer and fine speckle

Primer and fine speckles are usually applied before and after applying a PCC. Primers are applied on the concrete surface as well as the reinforcement. Thus, the material has to have a good bond to concrete as well as to steel surfaces. Both usually contain aggregates with a maximum grain size below 1 mm so that they can be applied in small thicknesses. The application is normally done with a brush or equivalent tools.

The application of a primer before applying a PCC cannot be considered an option for a durable protection, but the primer is an integral part of a repair system. The function of such a primer is to increase the adhesion of the PCC to the surface and to close the pores of the substrate in order to limit water suction into the substrate out of the fresh mortar.

The fine speckle is not a required part of a repair system and has mainly the function to produce a smooth surface, especially if mortars are applied by spraying.

5.3.7 Corrosion protection coating

Depending on the desired repair principle, so-called corrosion protection coatings have to be applied on the rebars before the concrete repair can be executed.

Corrosion protection coatings can be made out of epoxy resins or cement-based materials. Epoxy-based materials are not recommended because these systems only work as a passive corrosion protection, if applied entirely around the reinforcement, which cannot easily be ensured under on-site conditions (see Section 6.13.3).

Cement-based corrosion protection coatings are usually combined with the primer and can be used without any restrictions because these materials protect the steel actively by their own alkalinity and not by building up a barrier around the steel like epoxy-based systems (see Section 6.13.2).

5.4 CRACK FILLING

5.4.1 General

According to EN 1504-5, the filling of a crack fulfils one of the following goals:

- Force-transmitting filling of cracks (F)
- Ductile filling of cracks (D)
- Swelling fitted filling of cracks (S)

These goals cover the most common crack repair needs. It has to be noted that especially the force-transmitting and ductile fillings cannot be combined.

Regardless of the goal of a crack filling, the status of a crack is also important while selecting a proper filling material. The minimum thickness, the moisture state of the crack. and the daily movements—if applicable—all affect the material selection, which can be more or less generalised due to the properties of each material, as shown in the following sections. These properties are also indicated on the CE label of crack filling materials in order to enable a proper material selection (see (DIN) EN 1504-5:2006).

5.4.2 Epoxy EP

Epoxy resins are used for a force-transmitting filling of cracks with crack widths greater than 0.1 mm. Epoxy resins are two-component materials, which are usually solvent-free and feature a rather low viscosity in the range of about 150 to 400 mPa·s. The adhesion to dry concrete is always excellent. Depending on the type of epoxy resin, the moisture content of the crack can vary, as well as the minimum crack width and minimum and maximum working temperatures. The required wetness of the crack is depending on the type of EP. In cracks through which water runs usually PUR is used.

Epoxy resins are normally delivered in separate containers (usually two single containers), which feature sizes according to the exact stoichiometric ratio of the specific epoxy resin.

After mixing the single components of an epoxy resin according to the technical data sheets of the manufacturer, the pot life—time between mixing and the beginning of the hardening process—regulates the maximum working time.

Due to the excellent adhesion strength as well as the high tensile strength of epoxy resins in general, the injected crack features a higher tensile strength than the surrounding concrete. If a repaired structural element is overloaded, new cracks might occur besides the crack filled with EP.

5.4.3 Polyurethane PUR

Polyurethanes are reactive polymers that are used for ductile filling of cracks. PUR will form ductile foam within the crack. Polyurethanes can be used in wet cracks and even cracks with a permanent water flow. Depending on the type of PUR, one or two components are required for the injection. Single-component PUR forms to a foam with water; two-component PUR (isocyanate and polyol) will only form a stable and impermeable foam after mixing the two components together.

Depending on the status of the crack, often several crack treatments are required. If cracks with permanent water flow have to be closed, first a single-component polyurethane is injected in order to reduce the water flow through the crack. The result of this injection procedure is foam with a comparatively high porosity, which still lets water penetrate, but with a significantly lower flow rate. This foam forming usually only takes a few seconds. After this first injection a second injection has to be conducted in order to reduce the water flow to zero. This injection is done with two-component polyurethanes, which form a foam with a rather low porosity. Because these materials usually need a few hours to react, the first injection ensures that the material is not washed out of the crack during the application process.

5.4.4 Acrylic gels

Acrylic gels for swelling fitted of cracks should be used with care because the material may not ensure a proper corrosion protection in the injected cracks. These gels should be used to inject soil or non-reinforced structures to reduce the water transport toward a structure. If acrylic gels shall be used in contact with steel, a certificate stating an effective and durable corrosion protection is required.

5.4.5 Cement-based filling materials

Cement-based filling materials can be used if the load-bearing capacity of a structural element has to be restored and the crack flanks are wet. Cement-based filling materials contain high-grade cements with a specific surface up to 16,000 cm^3/g and up to 95 wt% of cement with a maximum grain size of 16 µm. These materials also contain a high amount of flow agents and stabilising polymers so that the emulsions are not separated during the injection process. So even though the composition seems like cement paste, the dry mixture of a cement-based filling material has to be prefabricated and cannot be mixed on site.

Cement-based injection materials also feature other advantages. The fire resistance of the repaired construction does not differ from the rest of the structure, which can be important as the cracks are located inside a building and the repaired areas are necessary for the load-bearing capacity of the structure. Furthermore, the alkalinity of cement-based injection materials provides an active corrosion protection to the steel reinforcement.

5.4.6 Application technology

While injecting a crack, the cracks should feature their maximum crack width, which is usually given at low temperatures. Due to the required hardening processes of all injection materials, the minimum working temperature according to EN 1504-5 specified for each material has to be observed especially while working at low temperatures. Generally the minimum working temperature of polymeric materials as well as cement-based materials is greater than 5°C. EN 1504-5 also specifies a maximum working temperature. Both data result in a temperature interval for which the material is designed to be processed. It has to be noted that the entire structure should not be cooler or hotter than the minimum or the maximum temperature of the injection material.

Usually cracks are filled by using so-called packers, which are either glued to the surface or drilled into the cracked concrete in a manner that the drill hole is crossing the crack—see Figure 5.14. The angle of the drill hole should be approximately 45°. These drill packers are used for high-pressure applications, while the glue packers are used for low-pressure applications. Glue packers are made out of steel, aluminium, or plastics and feature a plate and a thread that enables us to adapt a tube or a reservoir of injection material.

Drill packers feature a hollow and threaded tube and a rubber cuff. After the packer is inserted into the drill hole, the packer is tensed and the material can be injected through the hollow tube. Due to the rubber cuffs, the drill hole is sealed and the injection material cannot leak out of the hole. The packers are placed along the crack, and between each packer the crack is usually closed by using putty. The crack is injected from the bottom to the top, so that air can escape out of the crack and the crack can be completely filled with the injection material.

In order to treat an entire crack the two types of packers have to be placed in a specific order along the crack, as shown in Figures 5.15 and 5.16. It is important that the effective radii of the packers are known, and thus the packers are placed overlapping in order to avoid air entrapments.

The injection of a single packer is continued until the injection material is leaking out of the next packer. In order to achieve a high filling rate, each packer should be injected twice.

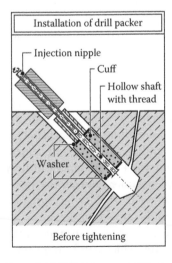
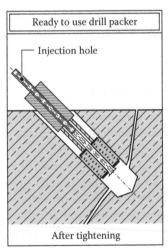
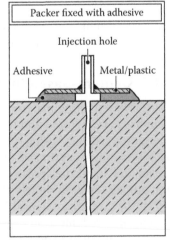

Figure 5.14 Different types of packers. Left and middle: Drill packer before and after placing in the borehole. Right: Packer glued to the surface.

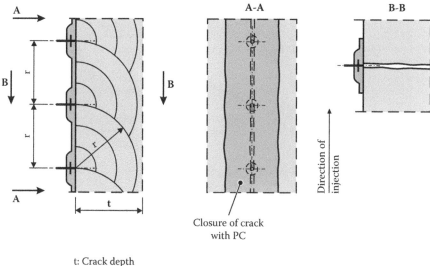

t: Crack depth
r: Effective radius of a single packer and
    distance of each packer (max. 60 cm)

Figure 5.15 Placing of glue packers along a crack according to RL-SIB 2001.

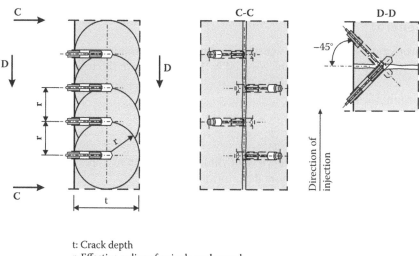

t: Crack depth
r: Effective radius of a single packer and
    distance of each packer (max. 60 cm)

Figure 5.16 Placing of drill packers along a crack according to RL-SIB 2001.

This has to be done during the pot life of the material. After finishing the injection, the packers as well as the putty are removed by using a grinder.

Using a low injection pressure over a longer time will lead to higher crack filling rates than high pressures combined with short injection times. In order to realise a low pressure over a longer time, packers with springs or balloons have been developed (see Figure 5.17). In both cases the pressure is applied without any additional energy, but only by a wounded spring or an extended balloon.

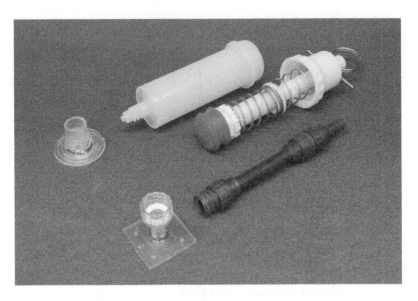

Figure 5.17 Spring (top) and balloon (bottom) type packers.

5.5 SURFACE PROTECTION SYSTEMS

5.5.1 General

Surface protection systems are a central element of concrete repair and rehabilitation. The types of surface protection systems covered by EN 1504-2 are the following:

- Hydrophobic impregnation
- Impregnation
- Coating

Table 5.2 gives a schematic overview of these three different types of surface protection systems in combination with a short description.

The standard EN 1504 does not cover any "flooring systems in buildings which are not intended to protect or reinstate the integrity of a concrete structure." These products are standardised in (DIN) EN 13813:2003. But the standard also states that "when products and systems complying with this standard are used in flooring applications that involve substantial mechanical loading, they shall also satisfy the requirements of (DIN) EN 13813:2003."

The following sections give an overview of the basics of the three different types of surface protection systems, as well as the most common materials used for each surface protection system.

5.5.2 Hydrophobic impregnations

Hydrophobic impregnations can be used for the protection of natural stone or concrete structures and feature the forming of an invisible film on the surface of the substrate while significantly reducing the water uptake of the substrate.

Table 5.2 Schematic drawing of surface protection system and short description

Type	Schematic drawing	Short description according to EN 1504-2
Hydrophobic Impregnation		• Water-repellent surface • Internal coating of the pores • No film on the surface • Little or no change of appearance
Impregnation		• Discontinuous, thin film on the concrete surface • Visible film • Pores are partially or totally filled
Coating		• Continuous protective layer on the surface • Visible film • Thickness varies typically between 0.1 and 5.0 mm

Hydrophobic treatments are usually silans or siloxans, which are diluted in water or alcohol. The materials are delivered as ready-to-use products, and on site the application can be done by spraying or brushing, depending on the viscosity of the material. The viscosity can be between water-like or creamy. All materials dry after application by evaporation of the dilution. While liquid systems dry off comparatively quick, creamy hydrophobic treatments remain longer on the surface. This usually results in a higher penetration depth, which is synonymous for a higher effectiveness, and thus a better and more effective protection of the substrate.

The hydrophobic treatment itself does not fill any pores, but only covers the pore walls. Due to the increase of the contact angle of the pore wall, water does not have the ability to penetrate into the substrate (see Figure 5.18). Still, water vapour or pressurised water can penetrate the pore system of the substrate.

The advantage of hydrophobic treatments is a water-repellent surface without changing the appearance as well as affecting the water vapour transport through the material. If hydrophobic treatments are used in combination with coatings, the main function of a hydrophobic treatment is to increase the long-term bonding strength of the coating to the surface.

Figure 5.18 Water drops on top of a concrete surface treated with hydrophobic agent.

5.5.3 Impregnations

Impregnations in the sense of EN 1504 are used to reduce the surface porosity in order to reduce the ingress of vapour or liquids and to increase the mechanical strength of the surface itself. Unlike hydrophobic treatments, the pores of the substrate are filled partially or fully and a discontinuous and thin film is formed on the surface of the substrate. Also, the appearance of the surface is altered by the application of an impregnation. Still, an impregnation does not feature a distinct layer thickness.

Generally impregnations are based on organic polymers, such as epoxies, polyurethane, or acrylic dispersions, which do not contain any fillers or pigments. The application is usually done by brushes, or if large horizontal areas have to be treated, the area is flooded and then the impregnation is spread with a slider (see Figure 5.19). If the impregnation is part of a surface protection system, the impregnation has to be gritted with sand in order to ensure a sufficient bond between the impregnation and the following layers.

5.5.4 Coating systems based on polymers

Coatings produce a continuous protective layer on the surface of concrete with a certain specified thickness. They are applied in order to rule out the ingress of harmful substances, to increase the mechanical resistance of concrete, and to bridge cracks—moving and non-moving. Typical thicknesses of coatings are between 0.1 and 5.0 mm, which might be enlarged due to special boundary conditions.

Depending on the intended use, coatings can be made, e.g., out of epoxy resins, polyurethanes, acrylates, polymer dispersions, or cement-polymer compounds. They might also contain aggregates, which are usually quartz sand with a grain size below 1 mm.

Coatings usually consists of more than one layer in order to meet the requirements given by EN 1504-5. Each layer features different functions, which are as follows:

- Impregnation (see above)
- Levelling layer
- Main protective layer
- Wear-out layer
- Top coat

Figure 5.19 Application of a surface protection system by hand.

This list of single layers is not defined in EN 1504-2. Exemplarily in the following paragraphs typical coating systems based on the German guideline for repair and rehabilitation of concrete (RL-SIB 2001) are described. The experience with surface protection systems has led to the knowledge that most of the requirements can only be achieved with more than one layer. So the following paragraphs briefly describe the function of each possible layer of a surface protection system.

Especially when the concrete surface is uneven and rough, the so-called levelling layer is applied. This layer has the following tasks:

- Levelling of the concrete surface
- Closing of pores, voids, and small defects of the concrete surface

This levelling layer is required because otherwise the next layers that represent the protective layers cannot be applied in a consistent thickness. The levelling layer is usually made out of polymer-modified cement mortar with a maximum grain size of 0.5 mm, and in exceptional cases 1 mm. The maximum thickness of the entire layer is about 3 mm. After applying the polymer-modified mortar on the surface, the surface is levelled so that peaks are not covered and the valleys of the surface are filled with mortar.

The main protective layer has one or more than one of the following features:

- Impermeable to water or harmful substances
- Impermeable to CO_2
- Crack bridging
- Resistivity to mechanical impact (abrasion or stresses caused by temperature changes)
- Resistivity to chemical attack

Analogous to the crack injection materials, surface protection systems can be classified into one of the following group:

- Sealing surface protection system
- Crack bridging surface protection system for dynamic crack openings
- Wear-out layers

Depending on the main feature, the materials that are used for the specific layers are selected. Generally for surface protections systems that should only seal a surface, stiff polymer resins, such as epoxy resins, are used to realise the layer. The thickness of such layers depends on the exposure of the layer. On vertical surfaces the layer thickness is approximately 0.3 mm; horizontal surfaces exposed to traffic require thicknesses around 5 mm. Due to the material properties, these layers not only seal the surface, but also work as a wear-out layer. In order to ensure a sufficient roughness, the layer is gritted with quartz sand or corundum, which has a higher resistance against abrasion than quartz sand (see Figure 5.20).

Figure 5.20 Application of sand to the freshly applied surface protection system.

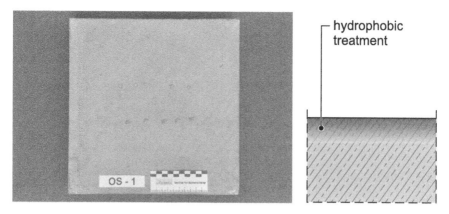

Figure 5.21 Left: Hydrophobic treatment on top of a concrete slab (upper side of the slab; lower side no hydrophobic treatment). Right: Schematic drawing.

If the surface protection system has to be crack bridging, flexible polymers such as poly-urethanes are used. Additional reinforcement layers made out of glass fibre textiles can be added to increase the resistivity of the crack bridging layer. The thickness of a main protective layer is usually between 1 and 4 mm. If these surface protection systems are exposed to traffic, the wear-out layer is also made out of polyurethane, but the material is not as flexible as the crack bridging layer. Also, these layers are highly filled with sand in order to ensure a sufficient durability.

In order to prevent the wear-out of the sand, a top coat has to be applied to the surface protection system. A top coat is usually made out of the same polymer as the protective layer and can be filled with pigments in order to meet optical requirements.

The layer thickness of a surface protection system represents one of the main material properties, which has to be specified by the manufacturer based on the requirements defined by EN 1504-2. The specified layer thickness is considered to be a minimum value that should not be fallen below to fulfil all necessary requirements selected by the designer.

The minimum layer thickness is usually specified as the thickness of the layer after drying and not during application. Due to the chemical hardening processes of the different polymers, the layer during the application is usually significantly thicker than the resulting layer after hardening. Thus, the technical data sheet of the manufacturer includes advice on how to calculate the resulting layer thickness based on the consumption during application.

Figures 5.21 to 5.27 show different types of surface protection systems according to the German guideline RL-SIB (RL-SIB 2001). The pictures show the real system as well as the schematic drawing of the entire system. The first system (OS-1) is a hydrophobic treatment, and the last one (OS-11) is a crack bridging surface protection system for parking decks. It can be seen that the increase of requirements leads to a significant increase of complexity and thickness of the entire surface protection system.

5.5.5 Coating systems based on bituminous materials

Polymer-based surface protection systems feature a small thickness of a couple millimetres resulting in a very low additional weight. Also, these surface protection systems are suitable for slow-moving traffic like in parking garages or similar. If a polymer-based surface protection system is exposed to fast-moving traffic or high-load traffic, such as truck traffic, mechanically more robust solutions are required.

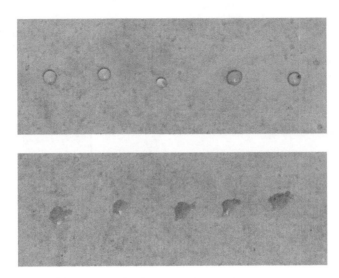

Figure 5.22 Detailed view of Figure 5.21. Top: Upper part with hydrophobic treatment applied on the concrete slab. Bottom: Lower part without hydrophobic treatment applied on the concrete slab.

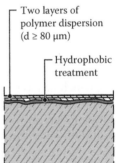

Figure 5.23 Left: Buildup of a surface protection system mainly for vertical surfaces on top of a concrete slab—concrete, hydrophobic treatment; two layers of polymer dispersion—not walkable, no crack bridging for corrosion prevention. Right: Schematic drawing.

For these application coatings based on bituminous materials, usually mastic asphalts are used. In Europe this type of asphalt is regulated in (DIN) EN 12970:2001 as well as (DIN) EN 13108:2006. If mastic asphalts should be used for ingress protection, not only the asphalt can be applied to the concrete surface, because the porosity of the mastic asphalt does not ensure a sufficient sealing and ingress protection.

Before applying the mastic asphalt, a polymer-based surface protection system or bituminous sheets have to be applied to the concrete surface. Both feature the sealant and the ingress protection. If polymers are used, the influence of the application of hot mastic asphalt has to be tested and the sealing effect cannot be influenced.

The mastic asphalt is then only the wear-off layer that protects the sealant. Usually for highway bridges, e.g., in Germany, the mastic asphalt is applied in two layers with a thickness

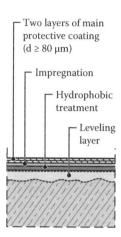

Figure 5.24 Left: Buildup of a surface protection system mainly for vertical surfaces on top of a concrete slab—concrete, levelling layer, hydrophobic treatment; base coat; two layers of main protective layer (polymer dispersion)—not walkable, no crack bridging for corrosion prevention and repair. Right: Schematic drawing.

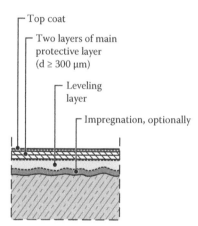

Figure 5.25 Left: Buildup of a surface protection system for surfaces on top of a concrete slab—concrete, impregnation, levelling layer, two layers of main protective layer (polymer dispersion); top coat—not walkable, crack bridging class B2 (—20°C) according to (DIN) EN 1062-7:2004. Right: Schematic drawing.

of 35 mm each. The two layers are necessary because the top layer is exposed to environmental loads and thus wears out slowly over time. If the status of the top layer requires a repair, the second layer protects the sealant so that the milling of the top layer does not harm it.

Figure 5.28 shows a typical application of such coatings based on bituminous materials on top of a bridge deck.

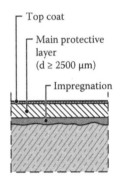

Figure 5.26 Left: Drivable surface protection for parking decks without crack bridging ability—concrete, impregnation, main protective layer (epoxy resin), top coat. Right: Schematic drawing.

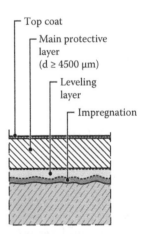

Figure 5.27 Left: Drivable surface protection for parking decks—concrete, impregnation, levelling layer, two layers of main protective layer (PUR), top coat—crack bridging class B3.2 (−20°C) according to (DIN) EN 1062-7:2004. Right: Schematic drawing.

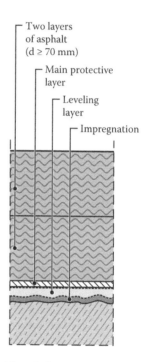

Figure 5.28 Left: Surface protection system applied under asphalt bridge decks—concrete, impregnation, levelling layer, main protective layer (epoxy resin), bituminous sealant, and asphalt. Right: Schematic drawing.

Chapter 6

Design principles and methods according to EN 1504

6.1 INTRODUCTION

As already described in Chapter 1, the design and selection of methods and products for protection and repair are based on a systematic classification scheme of principles and methods given in EN 1504-9. It contains a complete structured list of all principles to protect or repair concrete. For every principle the possible methods are given. Principles 1–6 are related to deterioration caused by corrosion of the concrete, and Principles 7–11 to deterioration caused by corrosion of the reinforcement.

Regarding specification of the methods, it is necessary not only to use the name of the method, like "coating" for repair works, but also the principle that is followed, e.g., a "coating for moisture control," or easier, "Method 2.3." This is why in the EN 1504 series the numbering system has been developed for all methods as listed in Table 6.1, which has been adapted in this book.

This systematics is a valuable basis for internationally standardised work. It would be appreciated if other countries outside of Europe also used this system to simplify international business in the field of repair and protection of concrete structures.

It requires some introductory training to understand and work with this system. To facilitate the first steps with this sophisticated system, it is described in detail in this chapter. For every method a schematic drawing and a summarising table with the most important information have been worked out, which are both not included in the standard EN 1504. Furthermore, examples and photographs from practice have been added.

Table 6.1 gives an overview of all principles and methods given in EN 1504.

The products for several methods are standardised in Parts 2–7 of EN 1504 as listed in Table 6.1. For the methods where no product standard is given in Table 6.1, other standards or recommendations outside of EN 1504 are relevant. More details on the requirements for the products are given in the following sections, where the methods are described.

For the methods of Principles 1–6 regarding corrosion of the concrete, the situation before application is shown in schematic figures covering three conditions: sound concrete, areas with bad quality of the concrete as represented by the gravel pocket in the centre, and a crack, as shown in the right half of the figures (see Figure 6.1).

6.2 PRINCIPLE I: PROTECTION AGAINST INGRESS

6.2.1 General

According to EN 1504-9 the approach of Principle 1 is to reduce or prevent the ingress of adverse agents, e.g., water, other liquids, vapour, gas, chemicals, or biological agents into the

Table 6.1 Principles and Methods Related to Defects in Concrete according to (DIN) EN 1504-9:2008–11

Principle	Examples of Methods Based on the Principles	Relevant Part of EN 1504
Principles and Methods Related to Defects in Concrete		
1. Protection against ingress	1.1 Hydrophobic impregnation	2
	1.2 Impregnation	2
	1.3 Coating	2
	1.4 Surface bandaging of cracks	
	1.5 Filling of cracks	5
	1.6 Transferring cracks into joints	
	1.7 Erecting external panels[a]	
	1.8 Applying panels[a]	
2. Moisture control	2.1 Hydrophobic impregnation	2
	2.2 Impregnation	2
	2.3 Coating	2
	2.4 Erecting external panels	
	2.5 Electrochemical treatment	
3. Concrete restoration	3.1 Hand-applied mortar	3
	3.2 Recasting with concrete or mortar	3
	3.3 Spraying concrete or mortar	3
	3.4 Replacing elements	
4. Structural strengthening	4.1 Adding or replacing embedded or external reinforcing bars	
	4.2 Adding reinforcement anchored in preformed or drilled holes	6
	4.3 Bonding plate reinforcement	4
	4.4 Adding mortar or concrete	3, 4
	4.5 Injecting cracks, voids, or interstices	5
	4.6 Filling cracks, voids, or interstices	5
	4.7 Prestressing (posttensioning)	
5. Increasing physical resistance	5.1 Coating	2
	5.2 Impregnation	2
	5.3 Adding mortar or concrete	3
6. Resistance to chemicals	6.1 Coating	2
	6.2 Impregnation	2
	6.3 Adding mortar or concrete	3
Principles and Methods Related to Reinforcement Corrosion		
7. Preserving or restoring passivity	7.1 Increasing cover with additional mortar or concrete	3
	7.2 Replacing contaminated or carbonated concrete	3
	7.3 Electrochemical realkalisation of carbonated concrete	
	7.4 Realkalisation of carbonated concrete by diffusion	
	7.5 Electrochemical chloride extraction	5
8. Increasing resistivity	8.1 Hydrophobic impregnation	2
	8.2 Impregnation	2
	8.3 Coating	2
9. Cathodic control	9.1 Limiting oxygen content (at the cathode) by saturation or surface coating	
10. Cathodic protection	10.1 Applying an electrical potential	

continued

Table 6.1 (continued) Principles and Methods Related to Defects in Concrete according to (DIN) EN 1504-9:2008–11

Principle	Examples of Methods Based on the Principles	Relevant Part of EN 1504
11. Control of anodic areas	11.1 Active coating of the reinforcement	7
	11.2 Barrier coating of the reinforcement	7
	11.3 Applying corrosion inhibitors in or to the concrete	7

ᵃ These methods may also be applicable to other principles.

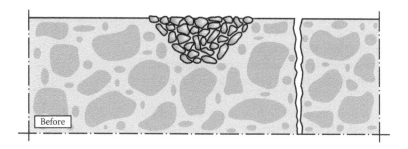

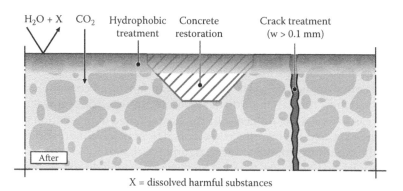

X = dissolved harmful substances

Figure 6.1 Schematic representation of Method 1.1 before and after application.

concrete. Typical adverse agents are CO_2, chlorides, or sulphates. Principle 1 is not related to chemical substances, which attack the concrete directly at the surface like acids. Increasing of the resistance to chemicals is addressed in Principle 6. Principle 1 can be achieved by eight different methods (see Table 6.1), which are described in the following sections.

6.2.2 Method 1.1: Hydrophobic impregnation

The ingress of water, including dissolved harmful substances, can be prevented by hydrophobic impregnation of the concrete. Figure 6.1 shows the application of this method schematically.

If a hydrophobic agent is applied, usually the gravel pockets have to be repaired and significant cracks have to be closed. It is recommended to close all cracks with a width of more than 0.1 mm. This is achieved by other methods, which will be described later. For repair of the gravel pocket Methods 3.1 to 3.3 could be applied, which belong to Principle 3 (concrete restoration). To close the cracks, Method 1.5 (closing of cracks) should be used. If considerable crack movements in the range of more than 0.1 to 0.2 mm are to be expected, it might

Table 6.2 Summary of Method 1.1

Method 1.1: Hydrophobic impregnation.

Principle 1: Protection against ingress.

Approach: Achieving a water-repellent concrete surface with low water uptake by application of a hydrophobic agent.

Typical applications: Vertical surfaces like façades, also horizontal surfaces; situations where the appearance of fair-faced concrete shall be preserved.

Special attention should be paid to:

- **Design:** Limited crack movements, unknown service life, etc.
- **Product requirements:** According to EN 1504-2.
- **Execution:** Careful surface preparation; concrete surface must be sufficiently dried out; high penetration depth shall be achieved.
- **Quality control:** Depth of penetration, hydrophobicity, etc.

Durability/maintenance: Regular inspections recommended.

Complementary methods: Usually Methods 1.5 and 3.1 to 3.3.

---

not be possible to prevent water ingress into the cracks by a hydrophobic impregnation. In this case, other methods like Method 1.3 (coating) should be preferred.

The products for hydrophobic impregnation are standardised in EN 1504-2. They can be applied by spraying, rolling, or temporary flooding. During application the concrete must be dried out to a certain extent to allow the hydrophobic agent to penetrate deep enough into the concrete. The durability of hydrophobic treatments is considerable, depending on the depth of penetration. Actual research shows that a penetration depth of several millimetres, usually in the range of 5 mm, is required to achieve long protection times, depending on the properties of the concrete and the hydrophobic agent.

Method 1.1 is often applied when the appearance of fair-faced concrete surfaces shall be preserved due to architectonical or aesthetical reasons. In these cases it is recommended to prepare trial areas to assess the influence of the hydrophobic impregnation on the appearance. Due to the expected drying effect, the concrete may look more light and uniform. Depending on the type of hydrophobic agent, significant changes also may occur. In the case of gels or pastes it has to be checked whether and when they will disappear after application.

Figure 6.1 shows schematically that after application of the hydrophobic agent, the concrete surface is protected against ingress of liquids. It should be noted that hydrophobic agents are open for gases, allowing water vapour to evaporate out of the concrete, but giving no protection against, e.g., ingress of CO_2. To reduce the rates of carbonation, usually coatings according to Method 1.3 are used.

On the contrary, a hydrophobic treatment can increase the rate of carbonation, because the mean humidity of the concrete decreases. As long as the concrete is dry, there is no problem. However, when accelerated carbonation has reached the reinforcement and the concrete gets wet due to loss of hydrophobicity in the course of time, the reinforcement can corrode. On the other side, if the structure is well maintained and inspections show that the effectiveness of the hydrophobicity decreases, the hydrophobic treatment should be renewed in time to keep the concrete sufficiently dry (Table 6.2).

6.2.3 Method 1.2: Impregnation

Impregnation aims to fill the pores of the concrete in the surface area to prevent any transport of liquids or gases through the concrete surface. Besides pore filling of the concrete, often additionally a thin film of the material for impregnation on the concrete surface is

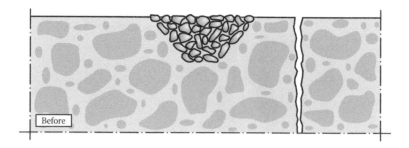

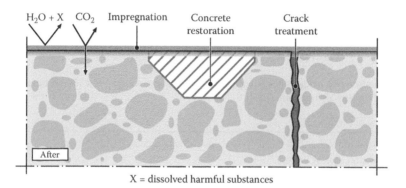

X = dissolved harmful substances

Figure 6.2 Schematic representation of Method 1.2 before and after application.

achieved. The result is sometimes also called pore blocking. Figure 6.2 shows the application of this method.

To allow a proper impregnation of the concrete surface, all areas with insufficient concrete quality have to be repaired. Here for repair of the concrete Methods 3.1 to 3.3 could be applied, which belong to Principle 3 (concrete restoration).

Furthermore, the cracks must be closed to ensure that the impregnation material cannot flow away through the crack. To close the cracks Method 1.5 (closing of cracks) should be applied.

The products used for impregnation are standardised in EN 1504-2. It should be noted that according to EN 1504-2 impregnations are not tested to be used for the reduction of the ingress of CO_2. To reduce the rates of carbonation coatings according to Method 1.3 are intended.

This method is normally used on floors and horizontal surfaces, where the impregnation material can easily be applied as shown in Figure 6.3.

One big disadvantage of impregnations is that they are not flexible. This means that if the crack width in concrete increases, e.g., in winter due to low temperatures, the impregnation will not be able to bridge the crack. Therefore, impregnations can only be used without additional protection when no crack movements are to be expected or open cracks are not critical.

However, in most cases in practice, crack movements cannot be excluded and following the nature of the principle of ingress protection, usually the cracks shall also be impregnated effectively. Therefore, this method is often used as a first layer of a coating system consisting of several different layers (see Section 5.5.4). In the example of Figure 6.3, after impregnation a sealing is applied followed by two bituminous protection layers.

Figure 6.3 Impregnation of a bridge deck using an epoxy resin according to EN 1504-2 before application of the sealing (see also Section 5.5.4).

Table 6.3 Summary of Method 1.2

Method 1.2: Impregnation.

Principle 1: Protection against ingress.

Approach: Closing the pores at the concrete surface (pore blocking).

Typical applications: Floors, horizontal surfaces.

Special attention should be paid to:

- **Design:** No protection when cracks move or new cracks occur, etc.
- **Product requirements:** According to EN 1504-2.
- **Execution:** Careful surface preparation; concrete surface must be sufficiently dried out.
- **Quality control:** Depth of penetration film thickness, etc.

Durability/maintenance: High durability, depending on the use.

Complementary methods: If required, Methods 1.5 and 3.1 or 3.2; note that impregnations can improve physical and chemical resistance (Methods 5.2 and 6.2).

Besides the effect of ingress protection, impregnations often also increase the physical resistance of the concrete and the resistance to chemicals. These additional improvements are described in Methods 5.2 and 6.2.

Investigations on the durability of impregnations based on epoxy resins of different parking decks in Germany have shown that they perform very well regarding chloride ingress, even 20 years after application. These tests have been carried out on old structures. Today for the protection of parking structures different types of coatings are used, but due to the problem of possible chloride ingress into the cracks, impregnations are not used alone anymore (Table 6.3).

6.2.4 Method 1.3: Coating

Coating of the concrete is one of the measures used most for protection and repair. Coatings can be simple paints, crack bridging, walkable, drivable, or even crack bridging and drivable, fulfilling highest requirements. In the sense of Method 1.3, coatings are used to protect the concrete against ingress of adverse agents. Compared to hydrophobic treatment or impregnation (Methods 1.1 and 1.2), coatings can also be used when significant crack movements are expected.

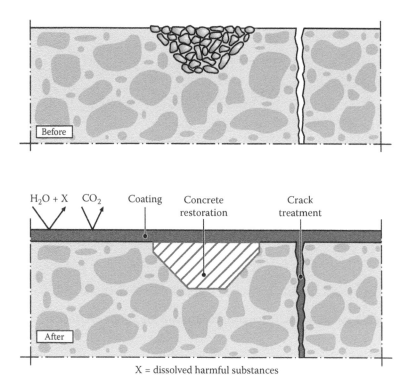

X = dissolved harmful substances

Figure 6.4 Schematic representation of Method 1.3 before and after application.

The coatings for Method 1.3 are standardised in EN 1504-2. It should be noted that in practice, usually coating systems consisting of several layers are used, which consist of products according to EN 1504-2. The thicknesses of the different layers and further required information for application have to be specified by the manufacturers (see Section 5.5.4).

Besides other performance criteria, the ability of the coating for crack bridging is important. The designer has to specify the expected maximum static and dynamic crack movements and select a product that is able to fulfil these requirements.

Figure 6.4 shows schematically the application of this method. Like for most other methods, areas with bad quality of the concrete, like gravel pockets, have to be repaired using Methods 3.1 to 3.3 (concrete restoration). To close the cracks, Method 1.5 (closing of cracks) should be used.

Normally coatings are applied on concrete surfaces all over (see Figure 6.5). However, if protection is not required all over, limited parts of a concrete surface can also be coated.

Figure 6.6 shows a parking structure where the coating has only been applied in areas where cracks are to be expected, i.e., the areas where tension stresses have to be expected on the top side of the deck. In this case the uncracked areas did not need protection against ingress of chlorides. However, this approach requires careful design, and the effect on the appearance has to be taken into account.

Method 1.3 is also often used to reduce the rate of carbonation to negligible values. This is an interesting preventive measure when the carbonation front has not reached the depth of the reinforcement, e.g., at façades. The carbonation resistance of the coatings as specified in EN 1504-2 is so high that the carbonation depth will more or less remain after application of the coating (carbonation brake) (Table 6.4).

Figure 6.5 Coating of a parking structure for protection against the ingress of chlorides.

Figure 6.6 Coating of subareas of a parking structure for protection against the ingress of chlorides in the area of cracks.

Table 6.4 Summary of Method 1.3

Method 1.3: Coating.

Principle 1: Protection against ingress.

Approach: Application of a coating that prevents ingress of adverse agents.

Typical applications: All types of concrete structures.

Special attention should be paid to:

- **Design:** Requirements have to be specified in detail (EN 1504-2).
- **Product requirements:** According to EN 1504-2.
- **Execution:** Careful surface preparation; concrete surface must have the required wetness; minimum thickness must be ensured.
- **Quality control:** Adhesion to concrete, coating thickness, etc.

Durability/maintenance: Inspections are recommended, depending on use.

Complementary methods: Usually Methods 1.5 and 3.1 to 3.3.

6.2.5 Method 1.4: Surface bandaging of cracks

Method 1.4 is one possibility to prevent the ingress of adverse agents into cracks in concrete. A flexible surface bandage is applied over a crack, protecting the crack like a subarea coating as used for Method 1.3. The application of Method 1.4 is shown in Figure 6.7.

The requirements for the products are not regulated in the product standards of EN 1504. Lots of different products are available, from coatings according to Method 1.3 to fibre- or mesh-reinforced multilayer bandage systems. Basic performance criteria are the ability to bridge certain crack movements and mechanical performance.

A special bandage has been used to close an old crack at the Cathedral of Aachen (Raupach et al. 2010). The designers of the repair works were looking for a flexible bandage with a service life that should be as high as possible to fulfil the requirements for the protection of cultural heritage. As a solution, a special bandage made of textile-reinforced concrete has been developed and applied. For the concrete a mortar has been used based on a mix design, which has already been used previously to repair parts of the historical masonry. As reinforcement, two meshes of carbon textiles impregnated with a special epoxy resin have been used. This bandage is technically watertight, because the crack widths in the bandage induced by opening of the crack to be protected will be significantly lower than 0.1 mm.

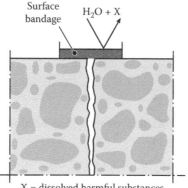

X = dissolved harmful substances

Figure 6.7 Schematic representation of Method 1.4 after application.

Figure 6.8 Crack in the Cathedral of Aachen, Germany, and installation of a bandage made of textile-reinforced concrete.

Figure 6.8 shows the installation of the bandage made of textile-reinforced concrete. The bandage has been applied quite widely over the crack to ensure not only that no water and adverse agents can penetrate through the bandage into the crack (Method 1.4), but also that the structure is strengthened to a certain extent to reduce the movements of the crack (Principle 4; see Section 6.5). It can be assumed that the service life of this bandage

Figure 6.8 (continued) Crack in the Cathedral of Aachen, Germany, and installation of a bandage made of textile-reinforced concrete.

is extraordinarily high. To monitor the proper functioning, strain sensors have been placed into the bandage. Details of this special bandage are given in Raupach et al. (2010).

Method 4.1 is usually applied for single cracks or extraordinary crack movements, which cannot be covered by coatings according to Method 1.3. If lots of cracks are present, a continuous full-surface coating according to Method 1.3. is more economic.

Especially if bandages have to bridge large crack movements and are exposed to high mechanical loads, a regular inspection is essential. In parking structures open cracks might lead to chloride-induced corrosion of the reinforcement after just one or a few winters, depending on the actual conditions (Table 6.5).

Table 6.5 Summary of Method 1.4

Method 1.4: Surface bandaging of cracks.
Principle 1: Protection against ingress.
Approach: Application of a flexible bandage over a crack, which prevents ingress of adverse agents.
Typical applications: Single cracks or cracks with large movements.
Special attention should be paid to:
- **Design:** Expected crack movements need to be specified, etc.
- **Product requirements:** Not covered in EN 1504; mainly crack bridging ability and mechanical resistance.
- **Execution:** Careful surface preparation, specified buildup.
- **Quality control:** Adhesion to concrete, bandage thickness, etc.
Durability/maintenance: Regular inspections recommended, depending on use.
Complementary methods: Method 3.1 for local defects in concrete.

6.2.6 Method 1.5: Filling of cracks

Method 1.5 is an alternative to Methods 1.4 and 1.6 to prevent ingress of adverse agents into cracks in concrete. The cracks are filled with a suitable material that remains in the crack and closes it. The aim is not to increase the load-bearing capacity in the crack, what is achieved by Methods 4.5 and 4.6 for Principle 4 (structural strengthening). On the contrary, Methods 4.5 and 4.6 often also lead to protection against ingress, so these methods may be used as a combination.

The application of this method in schematically shown in Figure 6.9. The crack is filled in a way that water and dissolved harmful substances cannot penetrate into or through the crack.

Generally, crack filling can be carried out with pressure by injection or without pressure, e.g., using a brush (see Section 5.4). Both methods can be used as long as the cracks are filled in a way that they remain closed after application.

The products for crack filling are standardised in EN 1504-5. There they are classified in injection products for force-transmitting, ductile, and swelling fitted filling of cracks. While force-transmitting filling belongs to Method 4.5, products for ductile or swelling fitted filling of cracks are able to ensure watertightness for specified water pressure classes. This means that the sealed cracks will also meet the requirements for ingress protection.

Figure 6.10 shows the injection of a crack with polyurethane using bore packers. Using this method the cracks can be filled over the whole depth.

The materials for ductile crack filling are to a certain extent flexible, allowing crack movements that are usually present in concrete structures. However, if the crack movements are too big, this method cannot be used.

For cracks in horizontal surfaces, like floors, alternatively crack filling without injection can be carried out using a brush or just by pouring the material (see Figure 6.11). To achieve a tight and ductile filling, the cracks should be opened on the top by saw cuts or milling. Usually the filling procedure has to be repeated several times until the crack is fully filled, because the filling material will slowly be soaked into the crack. This procedure can only be applied when the crack is dried to a sufficient depth to allow the crack filling material to penetrate deep enough into the crack.

The durability of this method generally depends very much on the real crack movements. If they are higher than expected, the filling material can lose bond on one or both sides from the concrete and ingress of aggressive substances again is possible. As already mentioned, especially in the case of chloride attack, the cracks shall not be open for longer periods of

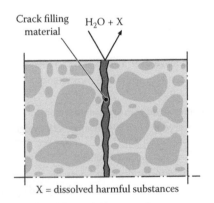

Crack filling material $H_2O + X$

X = dissolved harmful substances

Figure 6.9 Schematic representation of Method 1.5 after application.

Figure 6.10 Ductile filling of a crack by injection with polyurethane using bore packers as a possibility for Method 1.5.

Table 6.6 Summary of Method 1.5

Method 1.5: Filling of cracks.

Principle 1: Protection against ingress.

Approach: Filling the cracks with a sealing material.

Typical applications: All types of cracks.

Special attention should be paid to:

• **Design:** Method only applicable with limited crack movements, etc.

• **Product requirements:** According to EN 1504-5.

• **Execution:** Filling of the crack as completely as possible.

• **Quality control:** Drilling cores to check degree of filling, etc.

Durability/maintenance: Regular inspections recommended, depending on use.

Complementary methods: May be used combined with Methods 4.5 and 4.6.

time. Therefore, regular inspections are recommended depending on the conditions of use (Table 6.6).

6.2.7 Method 1.6: Transferring cracks into joints

Besides Methods 1.4 and 1.5, this is a third alternative to treat cracks with the aim to prevent the ingress of adverse agents. The crack is widened, e.g., by a saw cut and filled with sealing material using the well-known techniques to seal joints. This method is schematically shown in Figure 6.12.

It has to be ensured that the saw cut has no adverse influences on the structural behaviour and that the reinforcements are not cut. This could be a problem when the thickness of the concrete cover is very low.

The products used for this method are not covered in EN 1504. However, established systems for the sealing of joints are available for the different conditions of use. Special care has to be taken in the case of drivable joints (Table 6.7).

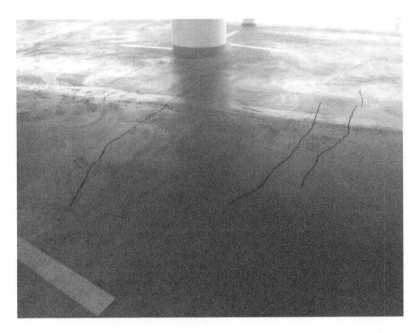

Figure 6.11 Cracks filled without pressure by milling to open the crack and pouring of ductile filling material as a possibility for Method 1.5.

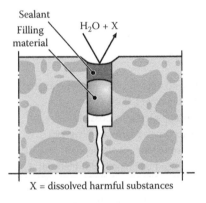

Figure 6.12 Schematic representation of Method 1.6 after application.

6.2.8 Method 1.7: Erecting external panels

Method 1.7 represents the protection against adverse agents by erecting external panels. Generally this method could be used for all types of concrete surfaces, but it can be very favourable at vertical surfaces like façades. The application of this method is shown schematically in Figure 6.13.

The external panel system, including joints and corner connections, must be able to prevent aggressive substances penetrating through. Such systems are available to protect against ingress of fluids like water, but not against gases like CO_2. To ensure watertight joints, welding can be used, e.g., for polymer or metal panels, but also other watertight solutions are available. The products for the panel systems are not standardised in the EN 1504 series.

Table 6.7 Summary of Method 1.6

Method 1.6: Transferring cracks into joints.

Principle 1: Protection against ingress.

Approach: The crack is widened and sealed like a joint.

Typical applications: Single cracks or cracks with large movements.

Special attention should be paid to:

- **Design:** Structural consequences of widening of the crack, etc.
- **Product requirements:** Regulations for sealing systems for joints.
- **Execution:** Specified buildup.
- **Quality control:** Adhesion of the sealant to both sides to the concrete, etc.

Durability/maintenance: Regular inspections and maintenance required.

Complementary methods: Method 3.1 for local defects in concrete.

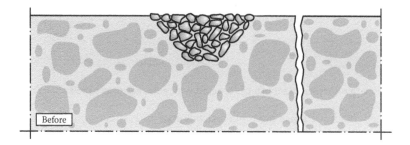

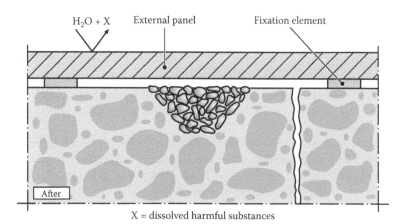

X = dissolved harmful substances

Figure 6.13 Schematic representation of Method 1.7 before and after application.

Figure 6.13 shows that areas with bad quality of concrete, like gravel pockets or cracks, can remain unrepaired when this method is used. However, it is recommended to repair these areas before erecting the panel system to increase the overall durability of the structure.

According to EN 1504-9 external panels are also used for Method 2.4 to achieve moisture control. The technical buildup of the panel system can be the same for both methods, but for Method 2.4 it has to be ensured that water from the concrete can evaporate through the panel system to allow a reduction of the water content of the concrete.

Figure 6.14 Basin in a sewage treatment plant protected with external glass panels.

External panels made of glass have been developed to protect concrete in an aggressive environment. It is already known that glass is watertight, technically impermeable for water vapour, and chemically resistant to most acids. Furthermore, the mechanical strength of glass has been improved substantially during recent years. This has led to the development of glass panels being glued to the concrete for use in aggressive environments like sewage systems and wastewater pipes, where concentrated biogenic sulphuric acid can develop. The glass panels are glued to the concrete using special mortars or flexible polymer-modified mortars.

Figure 6.14 shows external glass panels on the surface of a basin in a sewage treatment plant. It consists of sodium silicate glass as a float or toughened glass with a thickness of about 4–10 mm. The joints are sealed. A special flexible polymer-modified mortar is used as a connecting mortar.

Such glass panels have also been used in wastewater pipes, as shown in Figure 6.15. The glass panels have been prefabricated according to the inner radius of the tubes. For this application the smooth surface of the glass is advantageous, as it tends to stay clean and shows good hydraulic behaviour regarding water flow.

Another development for the protection of concrete with glass is thin glass foils with a thickness of only about 0.3 mm, which are bendable (see Figure 6.16). They consist of borosilicate glass produced as float glass. Usually they are installed overlapping. They are connected to the concrete using a special acid-resistant mortar.

Both glass systems have the potential to protect the concrete for decades and are especially interesting in cases where inspection, maintenance, and repair are difficult due to limited access (Table 6.8).

6.2.9 Method 1.8: Applying membranes

The protection of concrete against ingress of adverse agents by applying membranes is the objective of Method 1.8. On the contrary to panels according to Method 1.7, membranes

Figure 6.15 Wastewater pipe protected with external bendable glass panels.

Figure 6.16 Wastewater pipe protected with external bendable glass panels.

are not hard and stiff, but flexible and mostly ductile, e.g., like weldable bituminous linings or sheets.

The application of Method 1.8 is shown schematically in Figure 6.17. Concrete restoration has to be carried out to produce a suitable substrate for the membrane with sufficiently high surface tensile strength. Filling of the cracks is not generally necessary, because the

Table 6.8 Summary of Method 1.7

Method 1.7: Erecting external panels.
Principle 1: Protection against ingress.
Approach: Installation of a sealing panel system in front of a concrete surface.
Typical applications: Concrete exposed to aggressive substances.
Special attention should be paid to:
  • **Design:** Additional loads, solutions for details, etc.
  • **Product requirements:** Not in the EN 1504 series.
  • **Execution:** Specified buildup.
  • **Quality control:** Tightness, stability, etc.
Durability/maintenance: Inspections for leakages.
Complementary methods: Optionally Methods 1.5 and 3.1 to 3.3.

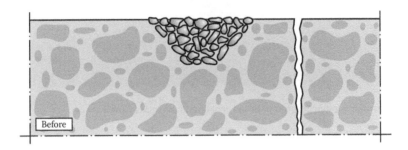

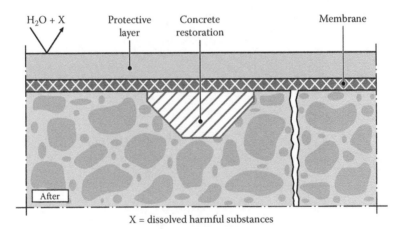

X = dissolved harmful substances

Figure 6.17 Schematic representation of Method 1.8 before and after application.

membranes are often able to bridge them. However, to get additional safety against ingress of adverse agents or to strengthen cracks according to Method 4.5 or 4.6, a crack treatment can also be useful.

As protection against ingress, membranes may also be used for Principles 2 (moisture control), 6 (resistance to chemicals), and 8 (increasing resistivity). However, the required performance, like evaporation capacity of water from the concrete, resistance against harmful chemicals attacking the structure, durability, etc., must be specified as the basis for the selection of products. The membranes themselves are not regulated in the EN 1504 series.

In the following examples of possible types of membranes for Method 1.8 are listed:

- Polymer foils
- Polymer linings or sheets
- Sprayable polymer foils
- Bituminous linings or sheets
- Polymer-modified bituminous linings or sheets, etc.

To increase the mechanical strength, membranes can be reinforced with fibres or meshes. As mechanical protection, often bituminous layers or other protective layers are applied onto the membranes, as schematically shown in Figure 6.17.

While sheets and linings are well known, the following example is a membrane made of a sprayable polyurethane foil. This technology offers the advantage that joints or overlapping areas are avoided and one continuous foil is generated. Sprayable membranes can easily be applied where the geometry of the surface to be protected is complex and irregular.

Figures 6.18 and 6.19 show exemplarily the repair and restructuring of a marketplace, which consists of a reinforced concrete construction with a large parking area below the marketplace. The use of de-icing salts has already led to a contamination of the upper surface of the concrete slab. To prevent further deterioration, the permeable paving has been removed and the concrete has been laid open.

For the new sealing the designer had to decide between sheets, linings, or sprayable systems. The surface was very irregular because it had joints, stairs, a fountain, openings for cables and de-watering, etc. Therefore, it was decided to use a sprayable system covered by a bituminous layer, a sand bed, and new pavement. As the spraying procedure is very quick, the whole surface could be handled at one go. After preparation of the surface by shot blasting, an impregnation according to Method 1.2 was applied before spraying of the membrane.

Figure 6.18 Spraying of a membrane according to Method 1.8 onto the concrete.

Figure 6.19 Continuous membrane after spraying (see Figure 6.18); afterwards a bituminous protection layer, followed by a sand bed and pavement, is applied.

Table 6.9 Summary of Method 1.8

Method 1.8: Applying membranes.

Principle 1: Protection against ingress.

Approach: A membrane is applied to the concrete surface, which prevents the ingress of adverse agents.

Typical applications: All types of concrete surfaces.

Special attention should be paid to:
  - **Design:** Crack movements, solutions for joints, protective layers, etc.
  - **Product requirements:** Not in the EN 1504 series.
  - **Execution:** Careful surface preparation; concrete surface must have the required wetness; minimum thickness must be ensured.
  - **Quality control**: Adhesion to concrete, membrane thickness, etc.

Durability/maintenance: Depending on system and use.

Complementary methods: Methods 3.1 to 3.3 for defects in concrete; optionally 1.5; for some systems 1.2 (impregnation) before application of the membrane.

This specific membrane has a dry thickness between 3 and 5 mm and remains ductile, and crack bridging even at temperatures of −20°C. After hardening, it is walkable, but for the intended use it requires a protective layer. In combination with the impregnation and bituminous protection layer, the new sealing will have a service life of decades (Table 6.9).

6.3 PRINCIPLE 2: MOISTURE CONTROL

6.3.1 General

The approach of Principle 2 is to adjust and maintain the moisture content in the concrete within a specified range of values to control adverse reactions. The concrete is allowed to dry, and moisture buildup is prevented. This method is often used to control alkali-silica reaction, sulphate attack, or freeze-thaw damage.

According to the informative annex of (DIN) EN 1504-9:2008-11, surface protection systems applied to vertical and soffit surfaces should be permeable to water vapour to allow moisture to escape from the concrete. Upper surfaces of horizontal concrete members, e.g., a suspended floor slab in a parking structure, may have an impermeable surface protection system applied. If the concrete contains extraordinarily high moisture, surface protection systems should not be applied.

To achieve moisture control five methods are available, which are described in the following sections. The first three methods, hydrophobic impregnation, impregnation, and coating, have already been introduced for Principle 1, and later it is shown that they are also used for Principle 8.

Regarding control of concrete corrosion, it should be noted that the drying effect of the concrete requires some time. Especially when the concrete is very wet, it may take some months or even years until the corrosion rates are sufficiently reduced to prevent damages. The fact that corrosion will continue for a certain time has to be taken into account for the design of the repair measure. If corrosion has proceeded so far that limit states like critical cracking will soon be reached, it might be too late to use moisture control, and alternative methods should be used that stop corrosion immediately.

6.3.2 Method 2.1: Hydrophobic impregnation

Moisture control can be reached by hydrophobic impregnation. For this method it is important to prevent ingress of water and to allow drying out by evaporation through the hydrophobic layer, as shown in Figure 6.20.

The products for hydrophobic impregnation are standardised in EN 1504-2. In this standard the drying rate coefficient describing the evaporation rate of water from the concrete through the hydrophobic layer, which is important for this method, is given as a performance criterion. When drying out is attempted to proceed quickly, the drying rate coefficient shall be close to 1, which means that evaporation is not hindered by the presence of the hydrophobic layer.

Regarding crack treatment, carbonation and appearance advice is given in Section 6.2.2 on Method 1.1. Besides these criteria, it has to be considered whether water may penetrate from the other sides to the concrete element, e.g., by capillary suction from the soil or other sources. If this cannot be excluded, other methods than 2.1 have to be selected.

To control concrete corrosion processes by moisture, it is necessary that the hydrophobic treatment is effective over the whole remaining service life. Therefore, inspections and maintenance are required. Most hydrophobic treatments can be repeated when the inspections show that the effectiveness is already reduced. When this method is intended for a long remaining service life, it is recommended to use such hydrophobic agents that can be applied repeatedly. This is not a performance criterion in EN 1504-2, but can be declared by the product manufacturer.

To determine the effectiveness of a hydrophobic impregnation, usually the water uptake of the concrete is measured after the treatment on site or at drilled cores. Actually research is performed on the durability of hydrophobic treatments (Antons et al. 2012). It shows that, e.g., mobile NMR (Nuclear Magnetic Resonance) is able to indicate the thickness of the hydrophobic layer at the concrete surface.

To determine the reduction of the water content within the concrete, sensors are available that can be embedded into the concrete. If they are placed in different depths or multiprobes are used, drying out can be monitored continuously (see Sections 6.10.4 and 8.4) (Table 6.10).

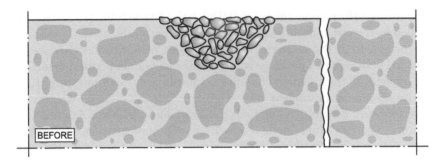

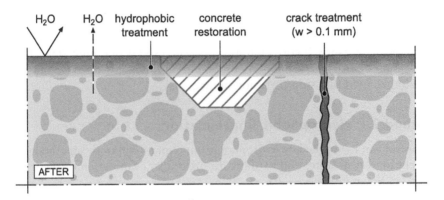

Figure 6.20 Schematic representation of Method 2.1 before and after application.

Table 6.10 Summary of Method 2.1

Method 2.1: Hydrophobic impregnation.

Principle 2: Moisture control.

Approach: Reduction of concrete corrosion rates by drying of the concrete.

Typical applications: Concrete corrosion like alkali-silica reaction, sulphate attack, and freeze-thaw in an early stage.

Special attention should be paid to:

- **Design:** As drying takes time, corrosion will continue after hydrophobic treatment and slow down slowly, etc.
- **Product requirements:** According to EN 1504-2.
- **Execution:** Careful surface preparation; concrete surface must be sufficiently dried out; high penetration depth shall be targeted.
- **Quality control:** Depth of penetration, hydrophobicity, etc.

Durability/maintenance: Regular inspections recommended.

Complementary methods: Usually Methods 1.5 and 3.1 to 3.3.

6.3.3 Method 2.2: Impregnation

Moisture control can also be achieved by impregnation of the concrete, which fills the pores in the area of the concrete surface. As preparation of the concrete surface, concrete restoration and crack filling have to be carried out if necessary, as shown in Figure 6.21.

As already explained in Section 6.2.3, impregnations are not crack bridging. As soon as new cracks develop in concrete or existing cracks get wider, the impregnation will crack and water can penetrate into the crack counteracting the principle of moisture control.

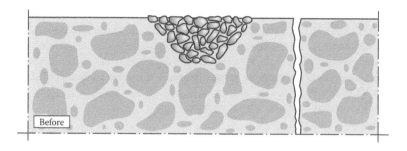

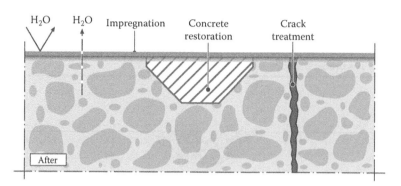

Figure 6.21 Schematic representation of Method 2.2 before and after application.

Therefore, it cannot be recommended to use Method 2.2 when crack movements or new cracks are expected. However, as Method 2.2 is predestined to be used on horizontal surfaces like floors, where often significant crack movements are expected, the fields of application of this method are limited. If only very few cracks are expected, it has to be considered to use Method 2.2 in combination with crack treatment, e.g., Method 1.5.

The products for impregnation are standardised in EN 1504-2. In this standard the drying rate coefficient is not used as a performance criterion like for the hydrophobic impregnation, but the permeability to water vapour. However, for impregnation products, like epoxy resins, it can be expected that the rate of drying out of the concrete is reduced considerably (Table 6.11).

6.3.4 Method 2.3: Coating

Coating systems can also be used for moisture control. Figure 6.22 shows schematically the application of this method. If necessary, as preparation of the concrete surface, concrete restoration and crack filling can be carried out. Compared to Methods 2.1 and 2.2, coatings have the advantage that crack bridging coatings are available. To allow moisture control, the coatings have to be impermeable for water from the outside and open for evaporation of water vapour from the concrete.

The coatings for Method 2.3 are standardised in EN 1504-2. The performance characteristics of the products for Method 2.3 are similar to those for Method 1.3 (protection against ingress), but there are no requirements regarding carbonation rate, ingress of chemicals, or chlorides because Method 2.3 is only focused on reduction of the water content of the concrete.

If drying of the concrete shall proceed quickly, coatings should be selected that allow high evaporation rates. As for all applications, the crack bridging performance should be specified carefully to prevent unexpected cracking (Table 6.12).

Table 6.11 Summary of Method 2.2

Method 2.2: Impregnation.

Principle 2: Moisture control.

Approach: Closing the pores of the concrete surface to reduce the water content of the concrete and the rates of concrete corrosion.

Typical applications: Limited; floors, horizontal surfaces.

Special attention should be paid to:

- **Design:** No protection when cracks move or new cracks occur; as drying takes time, corrosion rates will decrease slowly, etc.
- **Product requirements:** According to EN 1504-2.
- **Execution:** Careful surface preparation; concrete surface must be sufficiently dried out.
- **Quality control:** Depth of penetration, film thickness, etc.

Durability/maintenance: High durability, depending on use.

Complementary methods: If required, Methods 1.5 and 3.1 or 3.2.

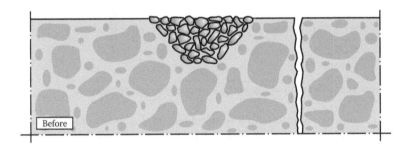

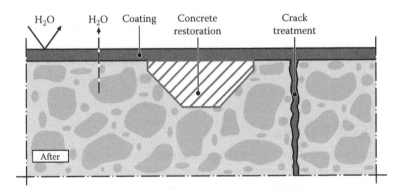

Figure 6.22 Schematic representation of Method 2.3 before and after application.

6.3.5 Method 2.4: Erecting external panels

External panels can be erected to reduce the water content of the concrete in front of concrete surfaces. The buildup looks similar to Method 1.7, but for Method 2.4 it is additionally important that the water from the concrete can evaporate through the panel system, as indicated in Figure 6.23.

Table 6.12 Summary of Method 2.3

Method 2.3: Coating.

Principle 2: Moisture control.

Approach: Application of a coating that prevents water ingress and allows evaporation of water from the concrete.

Typical applications: Concrete corrosion like alkali-silica reaction, sulphate attack, and freeze-thaw in an early stage.

Special attention should be paid to:

- **Design:** Requirements have to be specified according to EN 1504-2.
- **Product requirements:** According to EN 1504-2.
- **Execution:** Careful surface preparation; concrete surface must have the required wetness; minimum thickness must be ensured.
- **Quality control:** Adhesion to concrete, coating thickness, etc.

Durability/maintenance: Inspections are recommended, depending on use.

Complementary methods: Usually Methods 1.5 and 3.1 to 3.3.

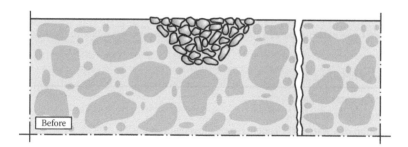

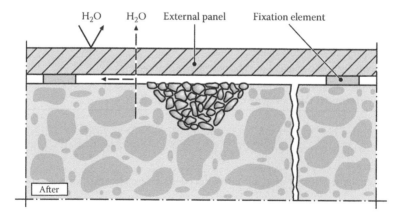

Figure 6.23 Schematic representation of Method 2.4 before and after application.

Figure 6.24 shows an example for an external panel system on a building façade. Such systems can also be used to improve the thermal insulation and appearance for the building. For applications on vertical surfaces like façades, the requirements of watertightness can be achieved much easier than for horizontal surfaces like floors. Rainwater will run down the panels, and the effective duration of the water exposure will be limited. For floors the

Figure 6.24 Example of an external panel system to control moisture according to Method 2.4.

Table 6.13 Summary of Method 2.4

Method 2.4: Erecting external panels.

Principle 2: Moisture control.

Approach: Installation of a watertight panel system allowing evaporation.

Typical applications: Concrete corrosion like alkali-silica reaction, sulphate attack, and freeze-thaw in an early stage, preferably at façades and roofs.

Special attention should be paid to:
  • **Design:** Additional loads, solutions for details, etc.
  • **Product requirements:** Not in the EN 1504 series.
  • **Execution:** Build up according to specifications.
  • **Quality control:** Tightness, stability, possibilities of evaporation, etc.

Durability/maintenance: Inspections for leakages.

Complementary methods: Optionally, Methods 1.5 and 3.1 to 3.3.

formation of puddles has to be expected, which extend the periods of water exposure and increase the performance requirements for the panel system, which shall additionally be able to allow evaporation.

With special cases it should also be considered to use external panel systems as protective roofs.

An external panel system can also be used for Principle 8, although it is not mentioned there as an extra method (Table 6.13).

6.3.6 Method 2.5: Electrochemical treatment

Method 2.5 is not specified in detail in the EN 1504 series. In recent years electrochemical methods for drying concrete, like electro-osmosis or the electro-osmotic pulse method, have been on the market for this method. However, there are serious doubts whether these methods are able to dry concrete significantly. As long as no scientific reports are available that present not only the assumed theory, but also the effect of these methods on the water content of the concrete quantitatively, these methods have to be treated carefully.

6.3.7 Method 2.6: Filling of cracks, voids, or interstices (not in EN 1504-9)

Filling of cracks, voices, or interstices can also be used for local moisture control to prevent damages due to corrosion of the concrete. This method is not included in EN 1504-9, but may be used in special cases, e.g., in combination with coating systems.

6.4 PRINCIPLE 3: CONCRETE RESTORATION

6.4.1 General

The approach of Principle 3 is to restore the original concrete of an element of the structure to the originally specified shape and function or to restore the concrete structure by replacing part of it.

Concrete restoration is normally carried out using hand-applied patch repairs, recasting with flowing concrete or mortar, or applying concrete or mortar by spraying.

Generally concrete restoration can be necessary for total surface areas or only parts of surfaces, as so-called patch repairs.

The strengthening of concrete structures, which can also be achieved by concreting, is not treated in Principle 3, but in Principle 4.

In the following sections the four methods of concrete restoration given in EN 1504-9 are described.

6.4.2 Method 3.1: Hand-applied mortar

Concrete restoration with hand-applied mortar can be used for the repair of relatively small areas. For larger areas, recasting according to Method 3.2 or spraying mortar according to Method 3.3 is technically and economically more favourable. The mortars are specified in EN 1504-3. As already mentioned, the goal of this method is only to replace the concrete with bad quality by new mortar or concrete, without strengthening the structure.

Figures 6.25 and 6.26 show examples of Method 3.1 applied to a column and the soffit of a slab. Single areas of the column obviously needed to be replaced with mortar. At the soffit of the slab shown in Figure 6.26, cracks have been widened so far that they could be closed with a suitable mortar to be applied overhead. This was possible because no crack movements were expected after repair.

It has to be decided whether the appearance of visible surfaces is important or not. Figure 6.27 shows as an example a big bridge pillar where lots of small defect areas were replaced with mortar. Obviously, there were no requirements regarding appearance of the patch repairs. Often the concrete surface is fully coated after patch repairs as a preventive measure to reduce the rate of carbonation. If concrete quality and thickness of the concrete cover are sufficient after the patch repairs, there is no technical need for a coating.

If this method is applied at fair-faced concrete façades, often additional requirements regarding appearance have to be fulfilled. This requires a systematic development of a mortar or concrete that fits to the existing concrete with respect to colour, glance, and structuring. An example for such a local repair of a gravel pocket is shown in Figure 6.28. The colour and structuration have been very well adjusted to the surrounding concrete.

Figure 6.25 Hand-applied mortar in single defect areas of a column.

Figure 6.26 Hand-applied mortar in widened areas of cracks in the soffit of an old slab, where no further crack movements are expected.

It should be noted that a perfect adjustment of the repair mortar to the appearance of the existing concrete is usually not possible, and that in the future differences cannot be excluded, even when using the same materials. Furthermore, it is difficult to achieve the same drying behaviour that leads to differences in the colours during drying. If a hydrophobic treatment is applied after local repair, the adjustment of colour only needs to be achieved for dry concrete. Anyhow, test areas should be located in time to show the expected appearance on site (Table 6.14).

Figure 6.27 Bridge pillar with numerous patch repairs according to Method 3.1.

Figure 6.28. Local repair of a concrete surface according to Method 3.1.

6.4.3 Method 3.2: Recasting with concrete or mortar

Recasting of defect areas with concrete or mortar is used as an alternative to applying concrete or mortar by hand or spraying. Generally, for recasting the same regulations can be applied as for new concrete construction. In any case, compatibility to the existing concrete and force transmitting through the transition zone between old and new concrete have to be considered.

Recasting is often carried out on the top of slabs where large areas of deteriorated concrete have been removed. Another common application is the filling of areas in vertical surfaces like walls or columns, where the concrete has to be replaced up to a considerable depth. In such cases, hand applying or spraying is not possible in one operation without waiting for a certain amount of time between the batches (Table 6.15).

Table 6.14 Summary of Method 3.1

Method 3.1: Hand-applied mortar.

Principle 3: Concrete restoration.

Approach: Replace defective concrete with mortar or concrete by hand.

Typical applications: All types of concrete surfaces.

Special attention should be paid to:
- **Design:** Requirements from appearance, etc.
- **Product requirements:** According to EN 1504-3.
- **Execution:** Surface preparation; complete removal of defective concrete.
- **Quality control:** Visual control, etc.

Durability/maintenance: —.

Complementary methods: Method 3.1 is often carried out prior to other methods.

Table 6.15 Summary of Method 3.2

Method 3.2: Recasting with concrete or mortar.

Principle 3: Concrete restoration.

Approach: Replace defective concrete with mortar or concrete by casting.

Typical applications: All types of concrete surfaces except soffits.

Special attention should be paid to:
- **Design:** Force transmission from old to new concrete, etc.
- **Product requirements:** According to EN 1504-3.
- **Execution:** Surface preparation; complete removal of defective concrete.
- **Quality control:** Visual control, etc.

Durability/maintenance: —.

Complementary methods: Often Method 1.8 (horizontal surface) or 1.3; Method 8.3 (vertical surface) or others, e.g., 1.1, 5.1, 6.1, 8.1.

6.4.4 Method 3.3: Spraying concrete or mortar

At vertical surfaces or soffits overhead, spraying concrete or mortar is an effective method. Due to the compaction effect by spraying, the quality of sprayed concrete or mortar is generally high. Further advice on sprayed concrete and mortar is given in (DIN) EN 14487-1:2006. Figure 6.29 shows exemplarily the procedure of spraying concrete.

Before spraying, it has to be ensured that the concrete substrate has a sufficient surface tensile strength. In most standards a minimum value of 1.0 N/mm^2 and a minimum mean value of 1.5 N/mm^2 are required (Table 6.16).

6.4.5 Method 3.4: Replacing elements

As EN 1504-9 includes all relevant methods for repair and protection of concrete, it also mentions replacing of elements. This may include materials other than reinforced concrete. Of course, the structural consequences during and after replacement have to be considered. This method is not further described within this book.

Figure 6.29 Photograph of the spraying procedure according to Method 3.3.

Table 6.16 Summary of Method 3.3

Method 3.2: Spraying concrete or mortar.
Principle 3: Concrete restoration.
Approach: Replace defective concrete with mortar or concrete by spraying.
Typical applications: Vertical surfaces and soffits of slabs or decks.
Special attention should be paid to:
   • **Design:** Force transmission from old to new concrete, etc.
   • **Product requirements:** According to EN 1504-3/EN 14487-1.
   • **Execution:** Surface preparation; complete removal of defective concrete.
   • **Quality control:** Visual control, etc.
Durability/maintenance: —.
Complementary methods: Often Methods 1.3, 8.3, or others, e.g., 1.1, 5.1, 6.1, 8.1.

6.5 PRINCIPLE 4: STRUCTURAL STRENGTHENING

6.5.1 General

According to EN 1504-9, the approach of Principle 4 is increasing or restoring the load-bearing capacity of an element of the concrete structure. It is essential when using Principle 4 that all stresses associated with repair and the original or deteriorated structure are considered. Certain systems may impose additional stresses on the repaired structure, resulting in changes of the original structural assumptions and behaviour.

While injection or sealing cracks (Methods 4.5 and 4.6) will not structurally strengthen a structure, injection may be used to restore the element to its structural condition prior to cracking, e.g., when temporary overloading has occurred.

Seven methods are available for Principle 4, which are explained in the following sections. Methods 4.1–3 describe ways to increase the load-bearing capacity by adding reinforcement, Method 4.4 to increase the load-bearing capacity in compression by adding mortar or concrete, Methods 4.5 and 6 to strengthen cracks, voids, or interstices, and Method 4.7 to strengthen a whole structural element by prestressing.

6.5.2 Method 4.1: Adding or replacing embedded or external reinforcing bars

According to Method 4.1 rebars are replaced, added into openings or slots into the concrete, or added outside of the existing structure into a new external concrete layer, as shown in Figure 6.30.

Replacement and adding of rebars within the existing concrete requires substantial concrete removal, e.g., by high-pressure water jetting. Figure 6.31 shows exemplarily the concrete of a building exposed to seawater, where the outer 20–30 cm of the concrete had to be removed due to corrosion, and the new reinforcement has already been installed into the designed position. To allow the transmission of forces to the old concrete, the remaining parts of the stirrups have been welded to the new reinforcement. After moulding, new concrete has been placed to renew the outer reinforced concrete zone.

Especially in a progressed state of corrosion, this procedure can also be applied to parking decks or bridges. However, as already mentioned, it has to be proven that the loads during and after these repair works can safely be adsorbed by the structure.

An example where external reinforcement has been installed outside of the existing structure is shown in Figure 6.32. Due to corrosion of the reinforcement, the slab had to be strengthened. The repair had to be carried out quickly and without disturbing the work in the factory buildings around. Therefore, a new reinforced slab had been installed onto the existing one without removal of the old concrete.

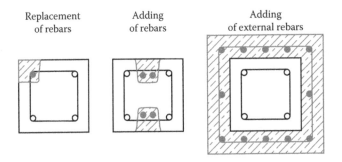

Replacement of rebars   Adding of rebars   Adding of external rebars

Figure 6.30 Schematic representation of Method 4.1 after application.

Figure 6.31 New reinforcement installed into a building exposed to seawater where the concrete had to be removed, before placement of the new concrete; see also Figure 6.83.

Figure 6.32 Adding of external reinforcement onto a concrete slab in a factory building.

Table 6.17 Summary of Method 4.1

Method 4.1: Adding or replacing embedded or external reinforcing bars.
Principle 4: Structural strengthening.
Approach: Adding or replacing damaged reinforcing bars.
Typical applications: Structures with insufficient load-bearing capacity, e.g., due to corrosion.
Special attention should be paid to:
  • **Design:** Structural analysis required.
  • **Product requirements:** Like for new construction.
  • **Execution:** Like for new construction.
  • **Quality control:** Like for new construction.
Durability/maintenance: —.
Complementary methods: Method 4.2.

To transmit the forces to the existing slab, steel anchors have been installed into the old concrete and connected to the new external reinforcement according to Method 4.2. Of course, this method is only possible when sufficient space is available and the additional loads can be taken over by the existing structure during and after strengthening (Table 6.17).

6.5.3 Method 4.2: Adding reinforcement anchored in preformed or drilled holes

As already mentioned in the previous section, it is necessary to connect new external reinforced concrete layers or elements to the existing concrete structure to transmit the forces between both parts. To achieve this, Method 4.2 can be used where reinforcement is anchored in preformed or drilled holes, as shown in Figure 6.33.

Preforming of the holes can be established, e.g., by chiselling or water jetting. Figures 6.34 and 6.35 show the situation of a bridge deck where the expansion joint had to be renewed.

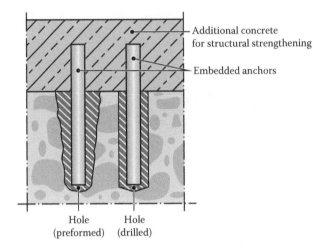

Figure 6.33 Schematic representation of Method 4.2 after application.

Figure 6.34 Expansion joint of a bridge after removing of the concrete.

The structural analysis showed that additional reinforcement was required in this part of the construction. This has been installed into drilled holes, as shown in Figure 6.35.

The products for this method are standardised in EN 1504-6. However, the procedure, including the materials and the requirements regarding preparation of the hole, distance to existing cracks, etc., is regulated in European or national approvals (Table 6.18).

6.5.4 Method 4.3: Bonding plate reinforcement

Strengthening can also be achieved by installation of bonding plates made of carbon fibre sheets or, in exceptional cases, by steel plates in concrete surfaces according to Method 4.3, as shown in Figure 6.36 schematically.

These bonding plates are glued to the concrete surface using special adhesive, which are standardised in EN 1504-4. However, the bonding plate systems, including plates and adhesives, are regulated in European or national approvals, where also details regarding surface

Figure 6.35 Detail of Figure 6.34. Additional reinforcement anchored in a drilled hole.

Table 6.18 Summary of Method 4.2

Method 4.2: Adding reinforcement anchored in preformed or drilled holes.
Principle 4: Structural strengthening.
Approach: Preparation of a hole, inserting a rebar and embedding with mortar.
Typical applications: Connections between new and old concrete elements.
Special attention should be paid to:
  • **Design:** Structural analysis required.
  • **Product requirements:** According to EN 1504-6 and approvals.
  • **Execution:** Careful preparation of the hole; complete filling with mortar.
  • **Quality control:** Pull-out tests, etc.
Durability/maintenance:
Complementary methods: Method 4.1 or 4.4.

preparation of the concrete, rules for the structural analysis, etc., are given. These approvals also specify to which percentage it is allowed to increase the load-bearing capacity of the structural element.

Today, mostly carbon sheets or wrappings with high tensile strengths are used for this method. Formerly, steel plates have been used, but they are difficult to install due to their high weight. Furthermore, their weight additionally has loaded the structures. As adhesives, usually epoxy-based materials are used.

Areas with bad quality of the concrete must be repaired and cracks filled. The concrete surface must have a sufficiently high tensile strength, concrete cover, and evenness as specified in the approvals of the specific systems. Special care has to be taken for the fire protection. Depending on the actual conditions, fire protection panels or other measures are required according to the approvals.

Generally, the bonding plates can be applied on the surface of the concrete or in slots, provided that sufficient concrete cover is available. Figure 6.37 shows as an example carbon sheets applied to a bridge soffit.

Especially for strengthening of columns, e.g., to increase the resistance to earthquake loads, large carbon sheets can be wrapped over the surface. The corners of square columns need to be rounded before application to prevent mechanical damages of the carbon sheets.

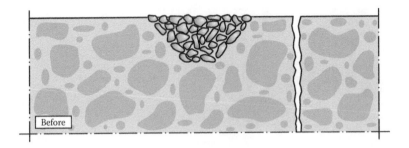

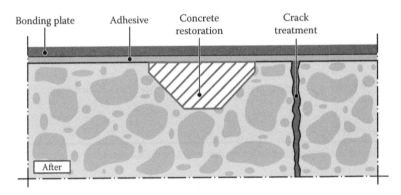

Figure 6.36 Schematic representation of Method 4.3 before and after application.

Figure 6.37 Strengthening of a bridge soffit using surface-mounted carbon sheets. (From Bepple, S., CFK-Sheets in Practise. In *Betoninstandsetzung heute und für die Zukunft. 11. Fachsymposium am,* Dortmund, March 25, 2003.)

Table 6.19 Summary of Method 4.3

Method 4.3: Bonding plate reinforcement.
Principle 4: Structural strengthening.
Approach: A high-strength sheet is glued to a concrete surface.
Typical applications: Structures with insufficient load-bearing capacity, e.g., due to corrosion.
Special attention should be paid to:
• **Design:** Structural analysis required, fire protection, etc.
• **Product requirements:** According to EN 1504-4 and approvals.
• **Execution:** Careful surface preparation of the concrete.
• **Quality control:** According to approvals.
Durability/maintenance: According to approvals.
Complementary methods: Methods 3.1 and 1.5 or 4.5 or 4.6.

To optimise the effect of strengthening, carbon sheets can be applied to structures in a prestressed condition. However, this procedure requires complex special equipment that controls the prestressing forces within the sheets during installation (Table 6.19).

6.5.5 Method 4.4: Adding mortar or concrete

Using Method 4.4, mortar or concrete is added to an existing concrete structure. The approach of this method is not to replace old concrete by new, but to place new concrete on top of the old concrete. As a consequence, the thickness of the structural element is increased. Consequently, the load-bearing capacity is also increased. However, it has to be checked that the increase of weight and of dimensions is no problem.

This method is schematically shown in Figure 6.38. To achieve a sufficient bond to the existing concrete, the areas with bad quality have to be removed and cracks need to be filled, e.g., using Method 1.5. It has to be ensured that the surface tensile strength of the concrete is higher than 1.5 N/mm^2 (mean value) and 1.0 N/mm^2 (lowest single value).

As this method is usually not applied to small surface areas, the concrete or mortar is directly cast or sprayed. Advice to these procedures is given in Sections 6.4.3 and 6.4.4 (Methods 3.2 and 3.3). For anchoring of the additional mortar or concrete, Method 4.2 can be used (Table 6.20).

6.5.6 Method 4.5: Injecting cracks, voids, or interstices

The standard method for strengthening concrete in the area of cracks, voids or interstices is injection with filling materials that are able to transfer loads like epoxy resins, cement-based mortars, or suspensions. On the contrary, polyurethanes or acrylic gels are used for sealing cracks, but not for strengthening. Method 4.5 is schematically shown in Figure 6.39.

Generally, it is intended to establish full transmission of ductile and tensile forces over the defect areas like cracks. This means that the injection material should be able to fill the crack as completely as possible and have a sufficient compressive, tensile, and adhesion strength in the concrete. Basic requirements of the injection materials are regulated in EN 1504-5 (see Section 5.4). In order to fulfil the requirements given by Method 4.5, usually epoxies or cement-based materials are used.

Most epoxy resins require a dry concrete surface to achieve sufficient bond and strength. However, during recent years special epoxy resins have been developed that harden very well also on wet concrete surfaces. Cement-based filling materials can of course be used on wet surfaces, but they require a certain minimum crack width to be able to penetrate

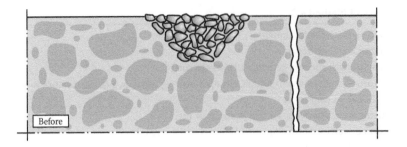

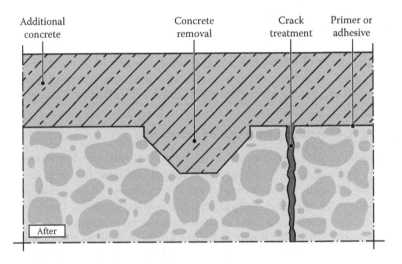

Figure 6.38 Schematic representation of Method 4.4 before and after application.

Table 6.20 Summary of Method 4.4

Method 4.4: Adding mortar or concrete.
Principle 4: Structural strengthening.
Approach: Increasing the thickness of a structural element.
Typical applications: All types of concrete structures.
Special attention should be paid to:
  • **Design:** Structural analysis required.
  • **Product requirements:** EN 1504-3 and –4 standards for new constructions.
  • **Execution:** Careful surface preparation.
  • **Quality control:** Adhesion, strength, etc.
Durability/maintenance: No special requirements.
Complementary methods: Method 1.5, 4.2, 4.5, or 4.6.

sufficiently deep into the crack. The application conditions like minimum crack width, wetness of the crack, etc., are given in EN 1504-5, and the information for use provided by the manufacturers.

Injection can be carried out using packers on top of the crack at the concrete surface or by drilling holes into the concrete crossing the crack about in the centre of the concrete, as shown in Figure 6.39. The injection parameters like pressure, etc., have to be set in a way that the water within the cracks is displaced and the cracks, voids, or interstices are filled as

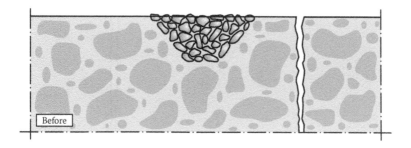

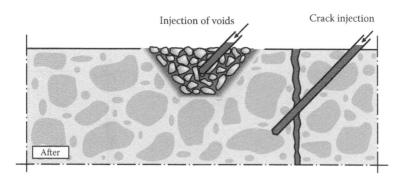

Figure 6.39 Schematic representation of Method 4.5 before and after application.

Table 6.21 Summary of Method 4.5

Method 4.5: Injecting cracks, voids, or interstices.

Principle 4: Structural strengthening.

Approach: Complete filling of defect areas with a hardening material by injection to achieve load bearing like in defect-free concrete.

Typical applications: Cracks, voids, or interstices in areas with high requirements to load-bearing capacity.

Special attention should be paid to:
- **Design:** Moisture condition in the crack, crack movements, etc.
- **Product requirements:** Basic requirements according to EN 1504-5.
- **Execution:** High degree of filling.
- **Quality control:** Degree of filling, etc.

Durability/maintenance: No special requirements.

Complementary methods: Method 3.1 or coating to improve appearance.

completely as possible with the injection material (see also Section 5.4). To check the degree of filling, cores can be drilled out of the injected area after hardening of the injection material (Table 6.21).

6.5.7 Method 4.6: Filling cracks, voids, or interstices

Method 4.5 deals with the filling of cracks, voids, or interstices by injection using a certain pressure. On the contrary, Method 4.6 means strengthening by filling defect areas without using pressure just by pouring. As already mentioned above, for strengthening purposes it is required to achieve an as complete as possible filling of the defect areas. However, without injection pressure, water that might be in the crack cannot be displaced significantly, and

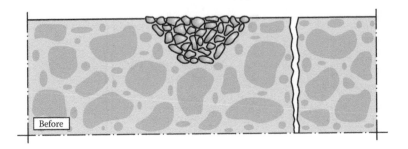

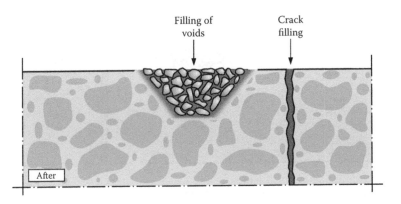

Figure 6.40 Schematic representation of Method 4.6 before and after application.

the penetration depth of the filling material is limited. A complete filling by pouring using epoxy resins may be possible when the crack is completely dry and has a certain width. If the concrete at both sides of the crack is only dry close to the concrete surface, a full filling of the crack probably cannot be reached. Using cement-based materials, the fields of application for this method are also limited to very wide and not water-saturated cracks.

Method 4.6 is shown schematically in Figure 6.40. As explained, the fields of application for this method are very limited. Therefore, it is not included in the German regulation for repair and protection of concrete structures. It is recommended to check the possible degree of filling in trials before deciding to use this method (Table 6.22).

Table 6.22 Summary of Method 4.6

Method 4.6: Filling cracks, voids, or interstices.

Principle 4: Structural strengthening.

Approach: Complete filling of defect areas with a hardening material by pouring to achieve load bearing like in defect-free concrete.

Typical applications: Wide, not water-saturated cracks.

Special attention should be paid to:

- **Design:** Trial applications to check possible degree of filling; alternative Method 4.5 (injection instead of pouring), etc.
- **Product requirements:** Basic requirements according to EN 1504-5.
- **Execution:** High degree of filling.
- **Quality control:** Degree of filling, etc.

Durability/maintenance: No special requirements.

Complementary methods: Method 3.1 or coating to improve appearance.

6.5.8 Method 4.7: Prestressing (posttensioning)

This method is not a simple intervention, but an extensive modification of the structural system. Using posttensioning by the installation of new prestressing steels into a concrete structure, the static system often can be changed in a way that the load-bearing capacity is considerably increased. This requires a complete recalculation and redesign of the structure, including the phases during the prestressing works. Typical fields of application are bridges, tanks, or supporting beams. The design aspects and materials for this method are covered by the standards for prestressed concrete.

6.6 PRINCIPLE 5: INCREASING PHYSICAL RESISTANCE

6.6.1 General

The approach of Principle 5 is increasing resistance to physical or mechanical attack.

According to EN 1504-9, this can be reached by three methods, which are described in the following sections.

6.6.2 Method 5.1: Coating

Method 5.1 represents coating of a concrete surface to increase its physical resistance, e.g., against abrasion or impact. It is shown schematically in Figure 6.41. As surface preparation, concrete with insufficient quality has to be replaced and cracks have to be closed.

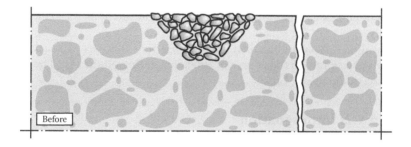

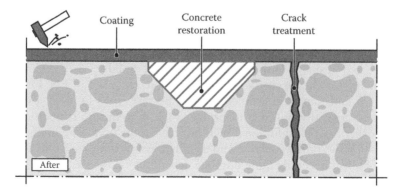

Figure 6.41 Schematic representation of Method 5.1 before and after application.

Table 6.23 Summary of Method 5.1

Method 5.1: Coating.
Principle 5: Increasing physical resistance.
Approach: Application of a coating to increase physical resistance.
Typical applications: Floors, surfaces exposed to abrasion or impact.
Special attention should be paid to:
  • **Design:** Crack movements or new cracks, mechanical loads, etc.
  • **Product requirements:** Basic requirements according to EN 1504-2.
  • **Execution:** Careful surface preparation, layer thickness.
  • **Quality control:** Adhesion strength, layer thickness, etc.
Durability/maintenance: Inspections recommended, depending on use.
Complementary methods: Methods 3.1 to 3.3, 1.5, 4.5, or 4.6.

To increase physical resistance of the concrete surface, coatings with high strength and good adhesion to the concrete are required, e.g., epoxy resins or similar polymers. Mortars and concretes for increasing physical resistance are assigned to Method 5.3.

The performance requirements for the coatings for Method 5.1 are given in EN 1504-2. They include abrasion resistance and impact resistance for all intended uses. Crack bridging is not generally postulated, but for certain intended uses it is specified by the designer.

For physical and mechanical resistance, small cracks reaching from the concrete through the coating are often acceptable, provided that the concrete quality is good enough to prevent breakouts at the boundaries of the crack. However, often cracks need to be sealed to prevent the ingress of adverse agents, e.g., chlorides. In such cases crack bridging coating systems could be a solution, but they need to be more ductile, what is counteracting to physical strength. Therefore, it has to be checked in cases where physical strength and protection against ingress into cracks are required which of the eight methods for Principle 1 shall be used additionally.

Regarding durability, it has to be taken into account that coatings exposed to abrasion or impact loads will deteriorate in the course of time, depending on the use. Therefore, regular inspections are recommended (Table 6.23).

6.6.3 Method 5.2: Impregnation

Alternatively to coating, an impregnation can be used to increase the physical resistance of a concrete surface. Method 5.2 is shown schematically in Figure 6.42. As surface preparation, concrete with insufficient quality has to be replaced and cracks have to be closed. It has to be considered to use the same material for an impregnation and to close gravel pockets or cracks, as shown in Figure 6.42. As explained for Method 1.2, impregnation materials penetrate into the pores of the concrete and may form a thin film on the concrete surface. Usually special epoxy resins are used as impregnation materials.

The requirements for the products for impregnation are given in EN 1504-2. Like for coatings, abrasion and impact resistance need to be proven to be used for this method. Impregnations are generally not able to bridge crack movements. If open cracks are not allowed due to other reasons, additional methods have to be used. The impregnation material hardens in the pores of the concrete and usually forms a thin, hard film on the concrete surface. Therefore, it is generally quite durable, depending on the use (Table 6.24).

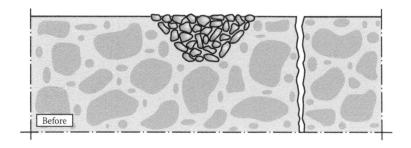

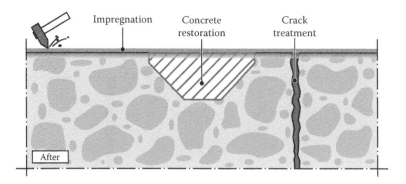

Figure 6.42 Schematic representation of Method 5.2 before and after application.

Table 6.24 Summary of Method 5.2

Method 5.2: Impregnation.
Principle 5: Increasing physical resistance.
Approach: Application of an impregnation to increase physical resistance.
Typical applications: Floors, surfaces exposed to abrasion or impact.
Special attention should be paid to:
  • **Design:** Crack movements or new cracks, mechanical loads, etc.
  • **Product requirements:** Basic requirements according to EN 1504-2.
  • **Execution:** Careful surface preparation.
  • **Quality control:** Adhesion strength, etc.
Durability/maintenance: Inspections recommended, depending on use.
Complementary methods: Methods 3.1 to 3.3, 1.5, 4.5, or 4.6.

6.6.4 Method 5.3: Adding mortar or concrete

A third possibility to increase the physical resistance is to add mortar or concrete to the concrete surface. Method 5.3 is schematically shown in Figure 6.43. As surface preparation, concrete with insufficient quality has to be replaced and cracks have to be closed. The surface tensile strength of the surface must be sufficient to achieve a durable adhesion between old concrete and the added layer of mortar or concrete.

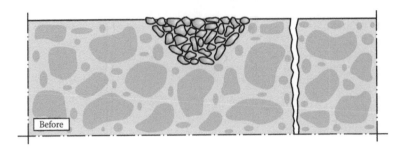

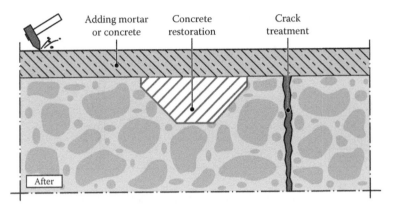

Figure 6.43 Schematic representation of Method 5.3 before and after application.

If the quality of the concrete at the surface is very low, even adding of a normal new concrete layer will increase physical resistance. If abrasion and impact resistance shall be increased to a high level, special mortars or concretes are available. The physical resistance can be increased by various measures, like using mixes with low porosity, special hard aggregates, fibres, or vacuum treatment, or a combination of all mentioned measures.

As this method has been included in EN 1504-9 in a very late stage, the mortars or concretes for Method 5.3 are not regulated in EN 1504-3. However, as mentioned, there are lots of mortars and concretes with high resistance to abrasion or impact on the market. To be able to select a suitable one, the parameters of use have to be specified as detailed as possible.

Regarding durability, again it has to be taken into account that surfaces exposed to abrasion or impact loads will deteriorate in the course of time, depending on the use. Therefore, regular inspections are recommended.

To increase the physical resistance of the added mortar or concrete, an impregnation according to Method 5.2 can be used on top of the added mortar or concrete (Table 6.25).

6.7 PRINCIPLE 6: INCREASING RESISTANCE TO CHEMICALS

6.7.1 General

The approach of Principle 6 is increasing the resistance of the concrete surface to deteriorations by chemical attack. The resistance of concrete to different classes of environmental attack is defined in EN 206. Severe chemical attack by chemicals is defined in EN 13529. The products used for the three methods to achieve this principle need to fulfil

Table 6.25 Summary of Method 5.3

Method 5.3: Adding mortar or concrete.
Principle 5: Increasing physical resistance.
Approach: Adding a layer of mortar or concrete to increase physical resistance of an existing concrete surface.
Typical applications: Floors, surfaces exposed to abrasion or impact.
Special attention should be paid to:
• **Design:** Crack movements or new cracks, mechanical loads, etc.
• **Product requirements:** Not in the EN 1504-series.
• **Execution:** Careful surface preparation.
• **Quality control:** Adhesion strength, thickness, etc.
Durability/maintenance: Inspections recommended, depending on use.
Complementary methods: Methods 3.1 to 3.3, 1.5, 4.5, 4.6, or 5.2.

the requirements of the performance characteristic "resistance to severe chemical attack," which is tested according to EN 13529.

6.7.2 Method 6.1: Coating

Method 6.1 represents coating of the concrete surface to increase the resistance to chemicals, which is shown schematically in Figure 6.44. As surface preparation, concrete with insufficient quality has to be replaced and cracks have to be closed.

As mentioned above, only coating systems can be used that fulfil the requirements of Method 6.1 given by EN 1504-2, including resistance to severe chemical attack.

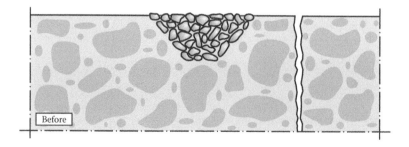

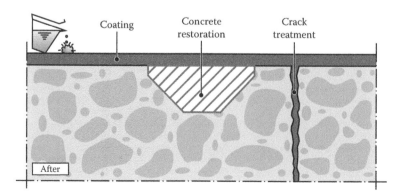

Figure 6.44 Schematic representation of Method 6.1 before and after application.

Table 6.26 Summary of Method 6.1

Method 6.1: Coating.
Principle 6: Increasing the resistance to chemicals.
Approach: Application of a coating to increase resistance to chemicals.
Typical applications: Surfaces exposed to severe chemical attack.
Special attention should be paid to:
- **Design:** Crack movements or new cracks, etc.
- **Product requirements:** Basic requirements according to EN 1504-2.
- **Execution:** Careful surface preparation.
- **Quality control:** Adhesion strength, layer thickness, etc.
Durability/maintenance: Inspections recommended, depending on use.
Complementary methods: Methods 3.1 to 3.3, 1.5.

For the selection of the coating system, the required ability to bridge crack movements has to be specified. Usually no open cracks are accepted in the case of severe chemical attack to the concrete (Table 6.26).

6.7.3 Method 6.2: Impregnation

Increasing of the resistance to chemicals can also be obtained by impregnation of the concrete according to Method 6.2, as shown in Figure 6.45. Again, a careful surface preparation is required; i.e., concrete with insufficient quality has to be replaced and cracks have to

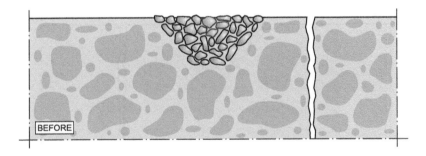

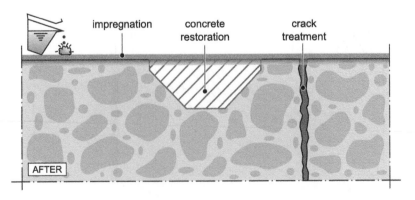

Figure 6.45 Schematic representation of Method 6.2 before and after application.

Table 6.27 Summary of Method 6.2

Method 6.2: Impregnation.
Principle 6: Increasing the resistance to chemicals.
Approach: Impregnation of the concrete surface to increase the resistance to chemicals.
Typical applications: Surfaces exposed to severe chemical attack, preferably in situations where no open cracks are expected.
Special attention should be paid to:
• **Design:** Crack movements or new cracks, etc.
• **Product requirements:** Basic requirements according to EN 1504-2.
• **Execution:** Careful surface preparation.
• **Quality control:** Adhesion strength, etc.
Durability/maintenance: Inspections recommended, depending on use.
Complementary methods: Methods 3.1 to 3.3, 1.5.

be closed. The materials should fulfil the requirements of Method 6.2 given in EN 1504-2, including resistance to severe chemical attack.

However, generally no open cracks are accepted in the case of severe chemical attack. Therefore, it has to be determined whether new cracks or crack movements are expected after impregnation. If this cannot be excluded, additional measures for the closing of the cracks are required, or other methods than impregnation have to be selected.

It has to be taken into account that the materials for closing the cracks also need to be resistant against severe chemical attack. Therefore, Method 6.2 is mainly interesting for cases where no open cracks are expected. In the other case, Method 6.1, using a coating with a crack bridging ability, is advantageous.

To enable a filling of the pores, the concrete must be dry near the surface; otherwise, the impregnation material cannot penetrate deep enough into the concrete (Table 6.27).

6.7.4 Method 6.3: Adding mortar or concrete

Method 6.3 stands for adding mortar or concrete to increase the resistance of chemicals. This is only possible when the mortar or concrete to be added has a significantly higher resistance to chemicals than the existing concrete. Figure 6.46 shows this method schematically.

Cement-based mortars and concretes generally are not chemically resistant, e.g., against most highly concentrated acids (see Section 2.2.3.2). Therefore, it clearly has to be specified against which chemicals the resistance shall be improved. Therefore, the fields of application of Method 6.3 are limited. However, for certain severe chemical attacks special high-performance concretes might be used that have been developed and tested to withstand the specified chemical attacks, e.g., according to EN 13529.

It has to be taken into account that mortars and concretes are not able to bridge crack movements. If open cracks cannot be excluded, additional measures are required to prevent problems in the areas of cracks, or Method 6.1 can be used (Table 6.28).

6.8 PRINCIPLES AND METHODS RELATED TO REINFORCEMENT CORROSION

6.8.1 General

Principles 7–11 and the corresponding methods are related to corrosion of the reinforcement. To demonstrate how the different methods can be applied, the schematic drawing shown in

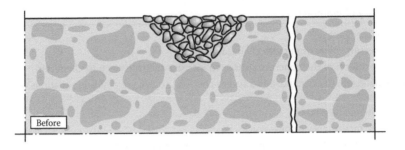

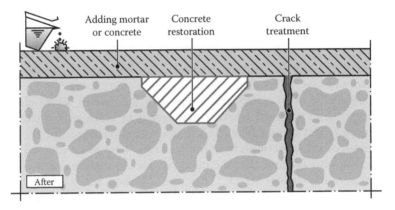

Figure 6.46 Schematic representation of Method 6.3 before and after application.

Table 6.28 Summary of Method 6.3

Method 6.3: Adding mortar or concrete.

Principle 6: Increasing the resistance to chemicals.

Approach: Adding a layer of mortar or concrete to increase the resistance of the concrete surface to chemicals.

Typical applications: Surfaces exposed to severe chemical attack, preferably in situations where no open cracks are expected.

Special attention should be paid to:

- **Design:** Crack movements or new cracks, etc.
- **Product requirements:** Not in the EN 1504 series.
- **Execution:** Careful surface preparation.
- **Quality control:** Adhesion strength, thickness, etc.

Durability/maintenance: Inspections recommended, depending on use.

Complementary methods: Methods 3.1 to 3.3, 1.5.

Figure 6.47 is used. It is divided into different zones representing areas with passive surface areas without corrosion (P) and zones that are depassivated due to different reasons (A–C).

Zone A has an insufficient concrete cover because the depth of the critical chloride content or carbonation x has already reached the concrete cover c_1. In practice, zone A is not necessarily small, but can cover the whole surface of a structural element.

Zone B is corroding due to insufficient quality of the concrete, as indicated by the symbol of the gravel pocket. Here the depth of the critical chloride content or carbonation x has already reached the deeper concrete cover c_2. This zone can also have any size in reality.

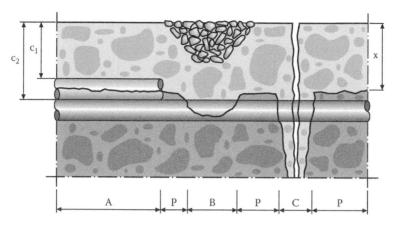

c_2, c_1 – Concrete cover

X – Depth of critical chloride content or carbonation

P – Passive surface area without corrosion

A – C – Areas with reinforcement corrosion due to

A – insufficient concrete cover

B – insufficient concrete quality

C – cracks

Figure 6.47 Schematic representation of typical situations regarding corrosion of the reinforcement used to explain the different methods for protection and repair in the following sections.

Zone C shows a crack crossing the reinforcement, through which chlorides or carbonation have penetrated quite deep into the concrete, as indicated by the route of the depth x.

In practice, often all three causes of corrosion (A–C) are present at the same structural element, but sometimes also only one or two of the three situations may be present. However, it is essential that before selecting any principles and methods, a detailed diagnosis has been carried out to know where which of the zones can be assumed.

In the following sections the principles and methods related to reinforcement corrosion are explained.

6.9 PRINCIPLE 7: PRESERVING OR RESTORING PASSIVITY

6.9.1 General

According to EN 1504-9, the approach of Principle 7 is creating chemical conditions in which the surface of the reinforcement is maintained in or is returned to a passive condition. This means that this method can be used as preventive protection before corrosion starts or for repair of already deteriorated reinforcement.

6.9.2 Method 7.1: Increasing cover with additional mortar or concrete

Method 7.1 is a preventive method: by increasing the concrete cover before the reinforcement starts to corrode, the initiation time of corrosion is prolonged, resulting in an extension of the remaining service lifetime. This method cannot be applied when depassivation has

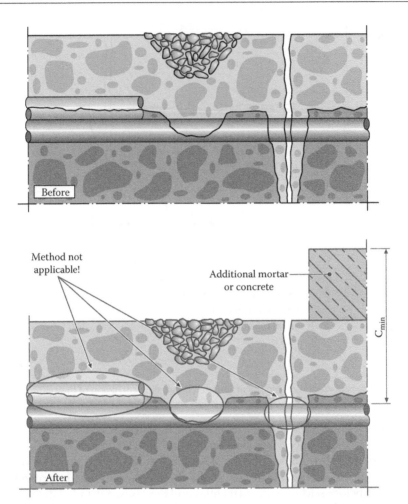

Figure 6.48 Schematic representation of Method 7.1 before and after application.

already occurred, e.g., by insufficient thickness or quality of the concrete cover or cracks, as shown schematically in Figure 6.48.

To prepare the concrete surface, the concrete with insufficient quality has to be replaced and cracks have to be closed. The thickness of the additional concrete layer provides an additional barrier against carbonation and chloride ingress. However, as the concrete may be already carbonated or contaminated with chlorides to a certain extent, this needs to be taken into account for the specification of the effective concrete cover. Furthermore, there are restrictions regarding the use of this method in the case of chloride contaminations.

The actual draft of the German guideline for maintenance of buildings (RL-SiB 2013) gives quantitative regulations for the use of this method. If the concrete is carbonated and not exposed to chlorides, Method 7.1 can be applied as far as the carbonation front has not reached the reinforcement, i.e., when a certain thickness X_0 of not carbonated concrete is available above the reinforcement, as shown in Figure 6.49. The effective concrete cover related to the remaining carbonation resistance $X_c(t_R)$ is the sum of the thickness of the added layer of mortar S_{min} of concrete and X_0. This equation can be used to specify the

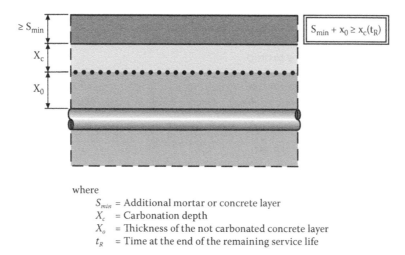

where

S_{min} = Additional mortar or concrete layer
X_c = Carbonation depth
X_0 = Thickness of the not carbonated concrete layer
t_R = Time at the end of the remaining service life

Figure 6.49 Schematic representation of Method 7.1 for carbonation-induced corrosion according to the actual draft of the German guideline for maintenance of concrete structures (RL-SiB 2013).

thickness S_{min} of the layer to be added depending on the design value of the remaining service life.

In the case of concrete exposed to chlorides, Method 7.1 can only be applied when there is a certain distance between the depth of the critical chloride content and the reinforcement, called X_0 in Figure 6.50. In the actual draft of the German guideline for maintenance (RL-SiB 2013) this thickness X_0 must be at least 10 mm. Furthermore, the chloride content within the remaining concrete shall be limited to 2 wt%/cem.

This is important because the chlorides will redistribute within the concrete after adding of the layer of mortar or concrete, and some chlorides will move to the direction of the reinforcement.

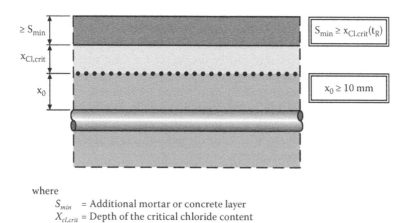

where

S_{min} = Additional mortar or concrete layer
$X_{cl,crit}$ = Depth of the critical chloride content
X_0 = Thickness of the concrete layer with uncritical chloride contents
t_R = Time at the end of the remaining service life

Figure 6.50 Schematic representation of Method 7.1 for chloride-induced corrosion according to the actual draft of the German guideline for maintenance of concrete structures (RL-SiB 2013).

Table 6.29 Summary of Method 7.1

Method 7.1: Increasing cover with additional mortar or concrete.

Principle 7: Preserving or restoring passivity.

Approach: A layer of mortar or concrete is added to increase the concrete cover as a barrier against further carbonation or chloride ingress.

Typical applications: Structures with insufficient concrete cover in a stage where the reinforcement is still passive.

Special attention should be paid to:
- **Design:** Redistribution of chlorides, etc.
- **Product requirements:** Basic requirements according to EN 1504-3.
- **Execution:** Careful surface preparation.
- **Quality control:** Adhesion strength, etc.

Durability/maintenance: No special requirements.

Complementary methods: Method 1.5.

To reduce the extent of redistribution of chlorides, the concrete surface should not extensively be wetted. Therefore, high-pressure water jetting should be avoided when the chloride contents of the concrete are already very high.

For dimensioning the thickness of the layer of mortar or concrete, the existing concrete cover should not be taken into account, but only the new layer with the thickness S_{min} (Table 6.29).

6.9.3 Method 7.2: Replacing contaminated or carbonated concrete

Method 7.2 is the traditional standard for the repair of reinforced concrete. All carbonated concrete or concrete with critical chloride contents is removed, the reinforcement is cleaned, and the breakout area is filled with concrete. The high pH value of the concrete passivates the reinforcement again and protects it against corrosion. This method is also called repassivation by alkaline mortar or concrete. It is shown schematically in Figure 6.51.

To improve the appearance and as protection against ingress of adverse agents, optionally a surface protection system according to Method 1.3 may be applied afterwards.

Generally, replacement of concrete can be necessary for total surface areas or only parts of surfaces, as so-called patch repairs.

In the case of local damages that shall be repaired by Method 7.2, the amount of concrete that needs to be removed has to be defined carefully. If not all deteriorated concrete (marked in Figure 6.52 as the depth x) is removed, corrosion will continue and can cause further damage. Figure 6.52 shows schematically possible corrosion mechanisms before and after local patch repairs (Raupach 2006).

If the reinforcement corrodes locally due to the low quality of the concrete, a corrosion element with anodes and cathodes will develop before repair. The reinforcement then corrodes in the anodic areas (a), while the cathodic areas (c) are to a certain extent cathodically protected by the action of the anode. This will lead to pronounced corrosion at the anode, but more or less no corrosion at the cathode. However, if the concrete in the area of the anode is removed, this cathodic protection gets lost.

In practice, sometimes the concrete is removed only partially, e.g., in weak or cracked areas. As a consequence, areas remain where the critical chloride content or carbonation

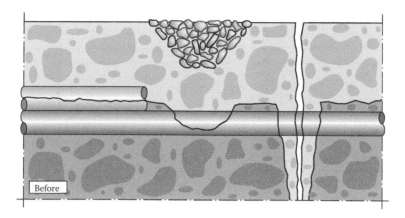

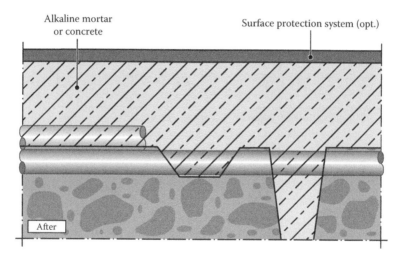

Figure 6.51 Schematic representation of Method 7.2 before and after application.

has reached the surface of the reinforcement, as indicated in Figure 6.52. In this situation corrosion may develop besides the patch-repaired area (a) because there is no further cathodic protection. These critical areas are known as incipient anodes. On the contrary, the reinforcement in the patch-repaired area can act as a cathode and accelerate corrosion. The mechanism is described in more detail in, e.g., Raupach (2006).

Such corrosion problems can only be avoided when all deteriorated concrete is removed where the critical chloride content or the depth of carbonation has reached the surface of the reinforcement. This means that the relevant criterion for the required removal of concrete is not the visible condition of the concrete at the surface, but the depth of the critical chloride content and carbonation front relative to the concrete cover. As a consequence, often concrete has to be removed that looks sound but is chloride contaminated or carbonated, to prevent further damage.

The actual draft of the German guideline for maintenance (RL-SiB 2013) gives quantitative details for the required depth of replacement X_{BA}, as shown in Figure 6.53. If the carbonation depth is extremely high in special cases, not all the carbonated concrete behind the

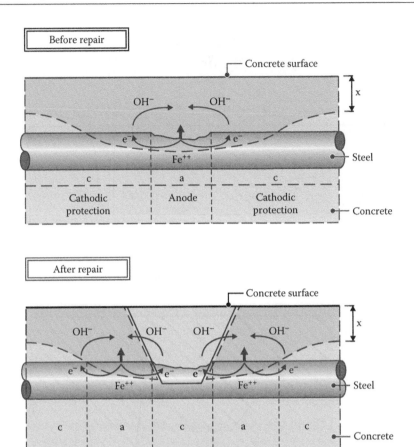

Figure 6.52 Possible corrosion mechanisms before and after patch repair. (From Raupach, M., *Cement and Concrete Composites* 28 (2006), no. 8, pp. 679–684, 2006.)

reinforcement needs to be removed, but within a distance of 10 mm when the steel diameter d_s is <16 mm or 15 mm, when d_s is thicker than 16 mm. This is necessary to allow a defect-free embedment of the reinforcement with mortar (see Figure 6.53).

If concrete is used instead of mortar, the distance behind the reinforcement where the concrete needs to be replaced has to be increased up to a value where defect-free embedment of the reinforcement is possible.

In the case of chloride-induced corrosion, all concrete needs to be replaced where the critical chloride content has been reached, as shown in Figure 6.54. This is recommended because chlorides may redistribute and cause corrosion problems.

If Method 7.2 is applied in the case of chloride-induced corrosion, this may lead to significant concrete breakouts. In such cases the load-bearing capacity of the structure without the concrete to be replaced needs to be proven before application of this method. This is especially important when concrete shall be removed at columns, beams, or other structural elements with high compression stresses. At structures with low load-bearing reserves, it can be necessary to temporarily strengthen the structure, e.g., by installation of extra columns or supports under slabs or beams.

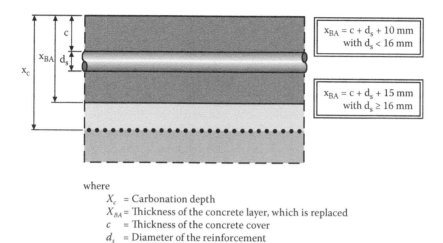

where

X_c = Carbonation depth
X_{BA} = Thickness of the concrete layer, which is replaced
c = Thickness of the concrete cover
d_s = Diameter of the reinforcement

Figure 6.53 Schematic representation of Method 7.2 for carbonation-induced corrosion according to the actual draft of the German guideline for maintenance of concrete structures (RL-SiB 2013).

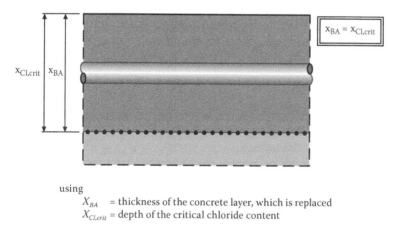

using

X_{BA} = thickness of the concrete layer, which is replaced
$X_{Cl,crit}$ = depth of the critical chloride content

Figure 6.54 Schematic representation of Method 7.2 for chloride-induced corrosion according to the actual draft of the German guideline for maintenance of concrete structures (RL-SiB 2013).

Figure 6.55 shows exemplarily a parking garage where chlorides have penetrated into open cracks for several years. This has led to very high chloride contents in the concrete along the cracks and subsequent corrosion of the reinforcement. As already explained in Section 2.3.5.1, the corrosion rates in the areas of cracks can be extremely high. Crack filling according to Method 1.5 would not solve the problem because the chlorides would remain and corrosion would continue. Therefore, the concrete with chloride contents above the critical limit value has been removed by high-pressure water jetting, as shown in Figure 6.55.

After water jetting the reinforcement is cleaned, and it can be determined how much steel has been removed by corrosion and whether additional reinforcement has to be added according to Method 4.1. A structural analysis of the structure is required to specify this.

After closing of the breakouts with (alkaline) concrete, it is recommended to apply a surface protection system according to Method 1.3 onto the whole concrete surface (Table 6.30).

Figure 6.55 Intermediate park deck after removal of chloride-contaminated concrete in the areas of cracks by high-pressure water jetting. (From Wolff, L., Bruns, M., Raupach, M. in *3. Kolloquium Erhaltung von Bauwerken*, ed. M. Raupach, Esslingen, January 22–23, 2013, pp. 211–221, Ostfildern: Technische Akademie Esslingen, 2013.)

Table 6.30 Summary of Method 7.2

Method 7.2: Replacing contaminated or carbonated concrete.

Principle 7: Preserving or restoring passivity.

Approach: Removing all carbonated and chloride-contaminated concrete and placement of alkaline concrete to repassivate the reinforcement.

Typical applications: Traditional method for all types of concrete structures.

Special attention should be paid to:

- **Design:** Structural analysis of the system or element, etc.
- **Product requirements:** Basic requirements according to EN 1504-3.
- **Execution:** Careful surface preparation.
- **Quality control:** Adhesion, strength, etc.

Durability/maintenance: No special requirements.

Complementary methods: Methods 4.1, 1.3 as protective coating system.

6.9.4 Method 7.3: Electrochemical realkalisation of carbonated concrete

Where the reinforcement is active or passive, additional corrosion protection can be provided by electrochemical realkalisation, which raises the alkalinity of carbonated concrete and thereby provides passivity to the reinforcement.

This method is described in CEN/TS 14038-1: *Electrochemical Realkalisation and Chloride Extraction Treatments for Reinforced Concrete—Part 1: Realkalisation*. This

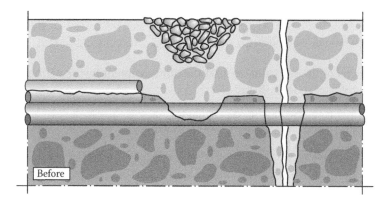

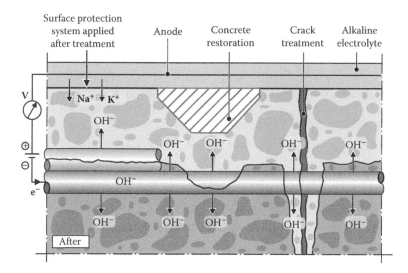

Figure 6.56 Schematic representation of Method 7.3 before and after application.

technical specification contains information on the principle, assessment, and repair of the structure, materials, and equipment, installation procedures, commissioning, operation and termination of treatment, final report, posttreatment coating, and monitoring.

The application of this method is schematically shown in Figure 6.56. For a certain time, usually several days, an anode is temporarily installed on the concrete surface, which is surrounded by an alkaline electrolyte. An electric DC is impressed between anode and reinforcement, creating an electrical field where all positively charged ions migrate toward the reinforcement. This increases the pH value of the concrete around the reinforcement.

To apply this method, concrete restoration and crack treatment may be necessary before, as shown in Figure 6.56. These repair measures should be carried out preferably using cement-based materials to allow a more or less homogenous current flow between anode and reinforcement during the treatment of this method. After removal of the anode and the electrolyte, the additional application of coatings according to Method 1.3 should be considered to prevent further carbonation and thereby extend the life of the treatment.

Several electrochemical processes start due to the impressed current, as described below.

At the reinforcement (cathode):

- Increasing of the pH value around the reinforcement by formation of hydroxyl ions due to oxygen reduction and electrolysis:

$$\frac{1}{2} O_2 + H_2O + 2\ e^- \rightarrow 2\ OH^-$$

$$2\ H_2O + 2\ e^- \rightarrow H_2 + 2\ OH^-$$

- Cathodic protection of the reinforcement during the period when the electrical current is impressed due to the significant reduction of the steel potential.

In the concrete:

Due to the applied electrical field, ion migration and humidity transport processes are induced:

- Negatively charged ions (Cl^-, SO_4^{2-}, CO_3^{2-}, HCO_3^-, OH^-) migrate from the reinforcement into the direction of the temporary anode.
- Positively charged ions (H^+, K^+, Na^+, Ca^{2+}, Fe^{2+}) migrate into the direction of the reinforcement.
- Humidity is transported to a certain extent into the direction of the reinforcement.

At the temporary anode:

- Development of oxygen by oxidation of hydroxyl ions and electrolysis of water, as well as acidification of the electrolyte by electrolysis:

$$2\ OH^- \rightarrow \frac{1}{2} O_2 + H_2O + 2\ e^-$$

$$H_2O \rightarrow \frac{1}{2} O_2 + 2H^+ + 2\ e^-$$

The material for the temporary anode needs to be electrochemically inert. Otherwise, it may corrode extensively and pollute the concrete surface with rust products, which are difficult to remove.

To demonstrate the effect of this method, laboratory tests have been carried out (Bruns et al. 2005). Figure 6.57 shows the setup used for the tests. The water/cement ratio of the concrete was 0.7 to allow quick carbonation and realkalisation. Specimens have been produced with CEM I and CEM III, concrete cover 2 and 3 cm, as well as a time of corrosion of 0, 3, and 6 months prior to the realkalisation.

The concrete slabs have been stored in a climatic chamber with 2 vol% CO_2 until the concrete has been completely carbonated from all sides. Then as an anode, a mesh of activated titanium has been installed on top of the specimen. As electrolyte, demineralised water has been used to investigate whether realkalisation is possible just by the processes at the reinforcement without external supply of alkalis.

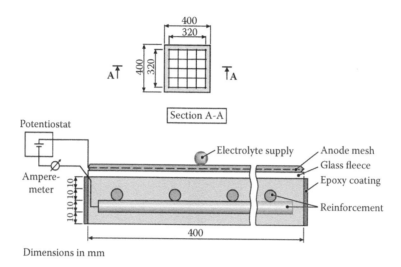

Dimensions in mm

Figure 6.57 Test setup for laboratory tests to demonstrate the effectiveness of Method 7.3.

As known from experience with Method 7.5, an intermittent application of the voltage increases the effectiveness of chloride extraction. Therefore, three variations have been investigated for the applied voltage for realkalisation:

- 40 V permanently
- 40 V for 12 h followed by 12 h without external voltage
- 10 V for 12 h followed by 12 h without external voltage

After seven days of application one specimen of each series has been opened to evaluate the thickness of the realkalised layer using the phenolphthalein method. As to be expected, Figure 6.58 shows that the thickness of the realkalised layer increases with the amount of current applied. However, there was no significant influence of the amount of voltage or intermittent application. To achieve realkalisation within short times, it is therefore advisable to use the highest voltages that are allowed to be applied at the specific building. Often this is roughly 40 V, which is used for the following tests.

The results shown in Figure 6.59 reveal that there is a quite high scatter, and that the influence of the duration of precorrosion and thickness of the concrete cover is not significant. However, there is a general trend that the thickness of the realkalised layer increases with the amount of applied current, as to be expected.

To achieve an adequate thickness of the realkalised layer of about 10 mm, the design value for the required amount of current should be increased from 200 Ah/m^2 according to CEN/TS 14038-1 to roughly 300 Ah/m^2 according to these results. However, the success of the treatment should be evaluated on site using the phenolphthalein test.

The durability of this method is limited by the effect that the concrete will continue to carbonate after the treatment. It has to be considered that the rate of carbonation of the realkalised concrete will be significantly higher than that of uncarbonated concrete. Therefore, often a coating is applied after the treatment as protection against recarbonation according to Method 1.3.

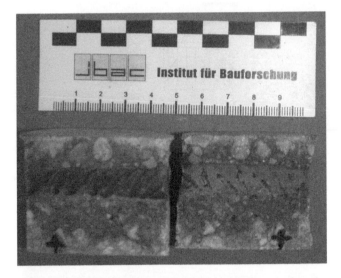

Figure 6.58 Thicknesses of the realkalised layer after treatment of seven days with different amounts of current densities related to the surface of the reinforcement (top: 33 Ah/m², bottom: 170 Ah/m²).

The effect of recarbonation of uncoated concrete has been shown in research (Tritthart and Baumgartner 2007). Figure 6.60 shows the concentration profiles of OH⁻ before, immediately after, and up to two years after electrochemical realkalisation of concrete specimens. It can be seen that the concentration of OH⁻ has been increased significantly in the depth of the reinforcement at 1.5 to 2.5 cm. However, already after six months most of the increased concentration of OH⁻ already disappeared, and after two years about the same conditions were found than before the treatment.

However, if a coating needs to be applied, it has to be considered whether application of a coating to increase resistivity according to Method 8.3 is already sufficient to solve the corrosion problem due to carbonation of the concrete. Furthermore, Method 7.3 is theoretically

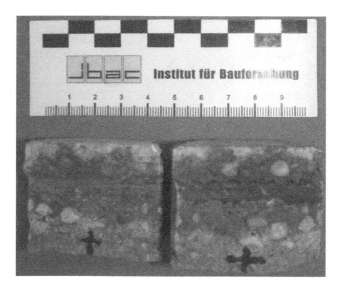

Figure 6.58 (continued) Thicknesses of the realkalised layer after treatment of seven days with different amounts of current densities related to the surface of the reinforcement (337 Ah/m²).

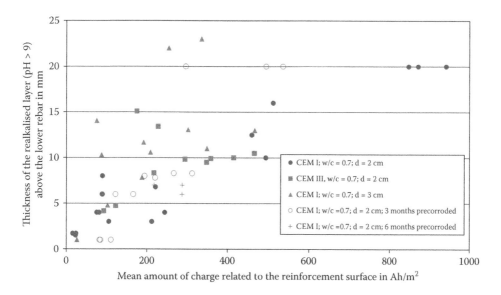

Figure 6.59 Thicknesses of the realkalised layer for different specimens depending on the amount of current applied (40 V permanently).

interesting to maintain fair-faced concrete, but if a coating is required after the treatment to prevent carbonation, this is no option.

So the fields of application of this method are very limited. It can be interesting to be used in combination with a coating system where a high reliability against corrosion is required, and where replacement of the carbonated concrete according to Method 7.2 is not intended (Table 6.31).

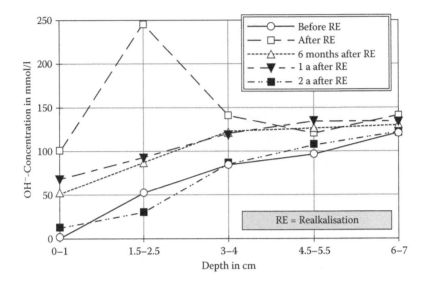

Figure 6.60 OH⁻ concentration profiles of concrete specimen before and after electrochemical realkalisation according to Tritthart (2007).

Table 6.31 Summary of Method 7.3

Method 7.3: Electrochemical realkalisation of carbonated concrete.

Principle 7: Preserving or restoring passivity.

Approach: Applying an impressed current to the reinforcement to increase the pH value of the concrete around the steel.

Typical applications: Limited.

Special attention should be paid to:
- **Design:** According to CEN/TS 14038-1; limited lifetime without coating.
- **Product requirements:** According to CEN/TS 14038-1.
- **Execution:** According to CEN/TS 14038-1.
- **Quality control:** According to CEN/TS 14038-1.

Durability/maintenance: Carbonation should be monitored after the treatment.

Complementary methods: Usually coating (Method 1.3).

6.9.5 Method 7.4: Realkalisation of carbonated concrete by diffusion

This method is not used in all European countries. One approach involves the application of a highly alkaline cementitious concrete or mortar to the surface of carbonated concrete, allowing the concrete to be realkalised through diffusion from the surface. The approach relies upon maintaining the concrete in a moist condition that permits effective diffusion to the depth of the reinforcing bar over the duration of the treatment, which can take many months. This approach has been used, e.g., in Germany, very successfully for more than 20 years.

The other approach involves application of an impermeable coating (e.g., made of polyurethane) to the concrete surface to keep the concrete behind the coating water saturated. Alkalis in the uncarbonated concrete will diffuse toward the coating over approximately one year to realkalise the concrete at the depth of the reinforcing bars. The concrete has to be saturated by groundwater, condensation water, etc., and cannot be exposed to frost. As the authors of this book do not have experience with this method and have no results of systematic research, they can actually not specify under which conditions an impermeable coating will provide realkalisation or not.

Regarding application of a highly alkaline cementitious concrete or mortar to achieve realkalisation by diffusion, systematic research has been carried out already more than 20 years ago (Haardt and Hilsdorf 1993). Based on the results, Method 7.4 has been implemented in the German guidelines for repair and protection of concrete structures since 1990. The application of this method is schematically shown in Figure 6.61. A layer of alkaline mortar or concrete is applied all over the concrete surface. After that, the previously carbonated layer of concrete is surrounded by alkaline concrete or mortar with a high pH value, and OH⁻ ions will diffuse into the carbonated concrete layer, increasing the pH value there until repassivation of the reinforcement is established.

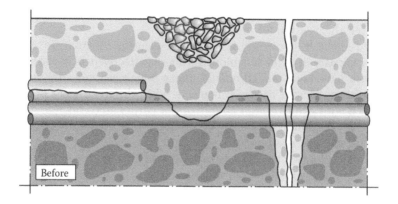

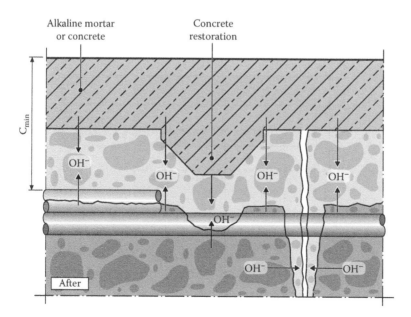

Figure 6.61 Schematic representation of Method 7.4 before and after application.

For a successful application of this method the following aspects need to be considered:

- The thickness of the layer of alkaline mortar or concrete must be so high that the carbonation front cannot pass through the new layer within the remaining service life. If this would happen, the realkalised layer would carbonate again very quickly due to its limited capacity against recarbonation.
- The distance between the depth of carbonation and the reinforcement must be limited. This is important to allow diffusion of OH⁻ ions from the noncarbonated to the carbonated old concrete, as shown in Figure 6.61.

Figure 6.62 shows how these requirements are regulated in the actual draft of the German guideline for maintenance of concrete structures (RL-SiB 2013).

Figure 6.62 shows that the minimum thickness of the layer of alkaline mortar or concrete S_{min} is 20 mm, or equivalent to the design value of carbonation depth at the end of the remaining service life. To be on the safe side, S_{min} should be estimated conservatively. Furthermore, Method 7.4 should only be applied when the depth of carbonation is less than 40 mm. However, this is given for most cases.

As mentioned above, this method is based on scientific research. In the following sections selected results are described, demonstrating the effectiveness of the method. Different types of concrete have been precarbonated under nonaccelerated, dry conditions, until a carbonation depth of 11 mm has been reached. Then different types of alkaline mortars have been applied with a thickness of 20 mm on the carbonated concrete surface. At the age of 14 days after application of the mortars (curing period: 3 days), the specimens have been stored in different climatic chambers.

The depth of realkalisation has been measured from time to time to allow a quantification of the effectiveness of this method, depending on the parameters, which have been varied, like humidity, concrete quality, and type of repair mortar.

Figure 6.63 shows the influence of humidity in the climatic chamber on the progress of realkalisation. As to be expected, the rate of realkalisation is increased with humidity. However, even in a quite dry environment at 65% relative humidity (rh) after 100 days, the

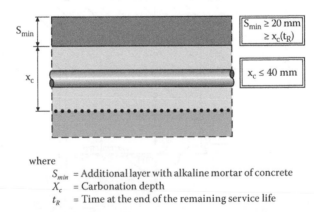

where

S_{min} = Additional layer with alkaline mortar of concrete
X_c = Carbonation depth
t_R = Time at the end of the remaining service life

Figure 6.62 Schematic representation of Method 7.4 for carbonation-induced corrosion according to the actual draft of the German guideline for maintenance of concrete structures (RL-SiB 2013).

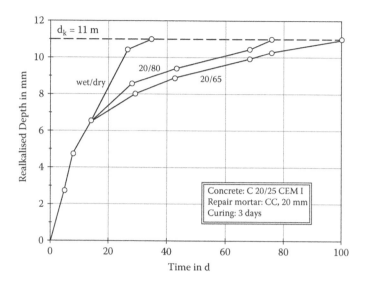

Figure 6.63 Influence of humidity on the progress of realkalisation.

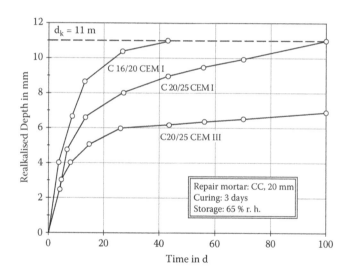

Figure 6.64 Influence of the type of concrete on the progress of realkalisation.

carbonated zone with a thickness of 11 mm is fully realkalised for the materials given in Figure 6.63.

Figure 6.64 shows the influence of the type of concrete. As expected, the rate of real-kalisation is low when the concrete has a low porosity. However, even at 65% rh and for CEM III, after 100 days a depth of 7 mm is reached and realkalisation is still progressing.

Figure 6.65 shows the influence of the type of repair mortar on the progress of realkalisa-tion. The two types of polymer-modified concretes, PCC I (5% polymer) and II (10% poly-mer), lead to low rates of realkalisation, because the amounts of polymers obviously inhibit the diffusion of alkalis. Therefore, it is recommended to use only cement-based mortars,

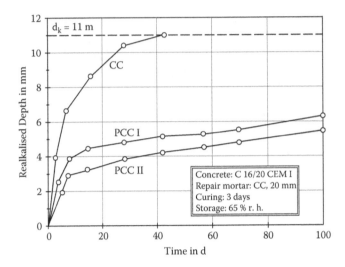

Figure 6.65 Influence of the type of repair mortar on the progress of realkalisation.

Table 6.32 Summary of Method 7.4

Method 7.4: Realkalisation of carbonated concrete by diffusion.

Principle 7: Preserving or restoring passivity.

Approach: Applying alkaline mortar or concrete on a carbonated concrete surface to increase the pH value of the carbonated zone by diffusion of OH^-.

Typical applications: All types of carbonated concrete structures.

Special attention should be paid to:

- **Design:** Sufficient thickness of the layer of mortar/concrete, etc.
- **Product requirements:** Cement based, high pH value.
- **Execution:** Achieve sufficient bond to the concrete.
- **Quality control:** Thickness, adhesion, etc.

Durability/maintenance: No special requirements.

Complementary methods: Usually not required.

preferably with CEM I, without polymers or other additions, which reduce alkalinity or the diffusion rate.

Corrosion tests have shown that after realkalisation the corrosion rates of the reinforcement decrease until the level of passive reinforcement is reached. These tests clearly indicate that Method 7.4 works with alkaline mortars or concretes. Experience from actual practice has shown that no corrosion problems have been detected when this method has been applied, as explained in Figure 6.62 (Table 6.32).

6.9.6 Method 7.5: Electrochemical chloride extraction

Where the reinforcement is passive or active due to chloride ion ingress, additional corrosion protection can be provided by electrochemical chloride extraction. This reduces the chloride ion concentration in the concrete surrounding the reinforcement and provides passivity, if sufficient chlorides are removed at the steel surface.

Guidance on this method can be obtained in Part 2 of CEN/TS 14038-2 regarding the principle, assessment, and repair of the structure, materials and equipment, installation

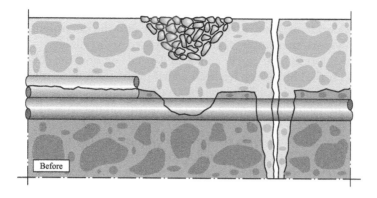

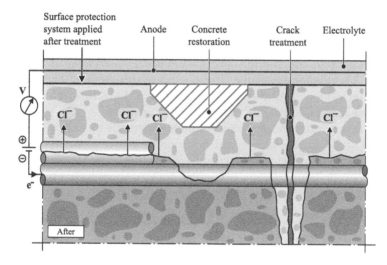

Figure 6.66 Schematic representation of Method 7.5 before and after application.

procedures, commissioning, operation and termination of the treatment, final report, post-treatment coating, as well as monitoring.

The application of this method is schematically shown in Figure 6.66. Like for Method 7.3 used to repair carbonated concrete, a temporary anode is installed that is surrounded by an alkaline electrolyte. Also, an electric DC is impressed between anode and reinforcement, creating an electrical field. However, the field is not applied to realkalise the concrete, but to induce migration of the chlorides away from the reinforcement out of the concrete toward the anode.

To apply this method, concrete restoration and crack treatment may be necessary, as shown in Figure 6.66. These repair measures should be carried out preferably using cement-based materials to allow a more or less homogenous current flow between anode and reinforcement during the treatment of this method. After removing the anode and the electrolyte, the additional application of coatings according to Method 1.3 should be considered to prevent further chloride ingress, and thereby extend the life of the treatment.

To reduce the chloride content sufficiently, usually several weeks are required. Similarly as for Method 7.3, often voltages of about 40 V are used. Experience from actual practice has shown that an intermittent application of the voltage can increase the rate of chloride extraction significantly.

As anode systems, generally the same materials can be used as for Method 7.3. It has to be considered that chlorides will enter from the concrete into the electrolyte around the anode. Again, modular systems that can be cleaned and reused are available.

The amount of chlorides that can be removed depends on how the chlorides have entered into the concrete. If the chlorides have been added to the fresh mix, what often is done for accelerated laboratory tests, usually about 50% of the chlorides can be removed. If the chlorides have penetrated into the hardened concrete, usually about 75% of the total amount of chlorides can be removed.

A total removal of all chlorides is not possible because of chloride binding and the effect that not only the chloride ions are activated to migrate out of the concrete, but also all other negatively charged ions. In the course of time most of the charge transport is established by OH^- ions that are produced by hydrolysis at the reinforcement, and the migration of chlorides decreases to negligible values.

Figure 6.67 shows chloride profiles at different areas of a parking deck measured before and after electrochemical chloride extraction. The profiles have been measured close to the reinforcement and between the reinforcements at two different locations, A and B.

Figure 6.67 shows that the chloride concentrations decreased from about >2.5 wt%/cem to <1 wt%/cem. This demonstrates the effectiveness of this method. However, a reduction to uncritical values < 0.5 wt%/cem could not be achieved. Therefore, it is uncertain whether the steel is repassivated or still actively corroding at low rates.

In such cases it is recommended to use additional methods to reduce the corrosion risk, like Methods 8.1 or 8.3. Furthermore, corrosion monitoring sensors could be used to observe the corrosion behaviour of the reinforcement (see Section 8.4). Simple potential mapping cannot be recommended immediately after application of the voltage because the reinforcement is polarised significantly, and even for corrosion specialists the measured potentials are difficult to interpret.

Figure 6.68 shows the electrochemical chloride extraction of the slab of a hollow box girder using a modular anode system where the electrolyte is placed in so-called pockets that can be cleaned and reused after the treatment (Schneck 2011). Cables are led from each module to a central computer that controls the voltages and monitors the individual electrical currents.

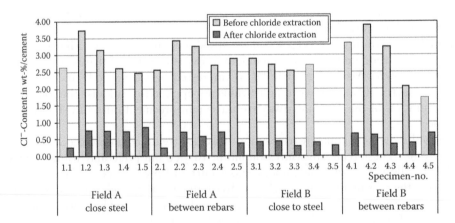

Figure 6.67 Chloride profiles measured before and after electrochemical chloride extraction of a parking deck.

Figure 6.68 Modular anode system for electrochemical chloride extraction. (From Schneck, U., in *Concrete Repair: A Practical Guide*, ed. M.G. Grantham, pp. 147–168, London: Routledge, Taylor & Francis Group, 2011.)

Figure 6.69 Electrochemical chloride extraction of the bottom part of a column in a parking structure during use. (From Schneck, U., in *Concrete Repair: A Practical Guide*, ed. M.G. Grantham, pp. 147–168, London: Routledge, Taylor & Francis Group, 2011.)

Figure 6.69 shows as an example the electrochemical chloride extraction of the bottom of a column in a parking structure where chlorides have penetrated from de-icing salts (Schneck 2011) during use of the structure. Traditional repair according to Method 7.2 would have been difficult during use. As shown in Figure 6.69, this method can be carried out without significant disturbances like noise or dirt.

An alternative method for electrochemical chloride extraction is cathodic protection (Method 10.1). However, this requires permanent electrical power supply and regular maintenance (Table 6.33).

Table 6.33 Summary of Method 7.5

Method 7.5: Electrochemical chloride extraction.

Principle 7: Preserving or restoring passivity.

Approach: Temporary application of an electrical field using an anode in an electrolyte to force the chlorides to migrate out of the concrete.

Typical applications: Where concrete removal is not possible and where no permanent electrical power can be applied, like for Method 10.1.

Special attention should be paid to:

- **Design:** According to CEN/TS 14038-2; if remaining chloride contents are too high, additional methods are required.
- **Product requirements:** According to CEN/TS 14038-2.
- **Execution:** According to CEN/TS 14038-2.
- **Quality control:** According to CEN/TS 14038-2.

Durability/maintenance: Remaining chloride content should be measured after the treatment; if it is too high, corrosion monitoring is recommended.

Complementary methods: Usually coating (Method 1.3).

6.9.7 Method 7.6: Filling of cracks, voids, or interstices (not in EN 1504)

Filling of cracks, voids, or interstices is used not only to protect against or repair concrete corrosion (Method 1.5), but also to preserve or restore passivity of the reinforcement. Therefore, the actual draft of the German recommendation for the maintenance of buildings (RL-SiB 2013) has introduced Method 7.6, which is actually not included in EN 1504.

If this method is applied, e.g., to fill young cracks in quite new buildings where the reinforcement is still passive, passivity can be preserved. For filling of the cracks all materials according to Method 1.5 can be used.

If this method is applied, e.g., to fill cracks where the reinforcement is already actively corroding, to restore passivity, cement-based crack filling materials could be used in the case of local carbonation-induced corrosion, because they are able to actively promote repassivation. However, this method can only be applied when the concrete between the cracks is not carbonated at the level of the reinforcement. To demonstrate the effect of repassivation in the areas of the cracks, electrochemical measurements like potential mapping are recommended.

In the case of chloride-induced corrosion in the area of cracks, this method is not suitable to restore passivity. As the success of this method depends on several factors, it is normally not used to restore, but to preserve passivity.

6.9.8 Method 7.7: Coating (not in EN 1504)

Coatings are used not only to protect against or repair concrete corrosion (Method 1.3), but also to preserve passivity of the reinforcement. Therefore, the actual draft of the German recommendation for the maintenance of buildings (RL-SiB 2013) has introduced Method 7.7, which is not included in EN 1504.

Coatings are often used to prevent carbonation or chloride ingress into the concrete to preserve passivity. In EN 1504 such coatings belong to Method 1.3. However, if it is distinguished consequently between corrosion of the concrete and corrosion of the reinforcement, this method belongs to Principle 7.

6.9.9 Method 7.8: Surface bandaging of cracks (not in EN 1504)

Surface bandaging of cracks is used not only to protect against or repair concrete corrosion (Method 1.4), but also to preserve passivity of the reinforcement. Therefore, the actual draft of the German recommendation for the maintenance of buildings (RL-SiB 2013) has introduced Method 7.8, which is not included in EN 1504.

6.10 PRINCIPLE 8: INCREASING RESISTIVITY

6.10.1 General

According to EN 1504-9 the approach of Principle 8 is increasing the electrical resistivity of the concrete up to a level where the corrosion rate of the reinforcement is negligible. As explained in the section on fundamentals of corrosion, no damages by reinforcement corrosion are expected when the concrete is dried out to an extent that is in the range for typical indoor situations, even if the concrete is carbonated down to the level of the reinforcement.

Compared to the situation of carbonated concrete, this principle is more difficult to achieve for chloride-contaminated concrete or, in the case of high chloride contents, even impossible. This is because the chlorides are hydrophilic and impede drying out. There is no accepted limit value for the range of chloride contents at which this principle can be used or not. The designer has to decide whether this principle shall be used in the case of chloride-induced corrosion. The risk of unexpected damages can be reduced by careful inspections or the use of suitable corrosion monitoring systems.

Regarding control of concrete corrosion processes by this principle, it has to be noted that the drying effect of the concrete requires some time. Especially when the concrete is very wet, it may take some months or even years until the corrosion rates are sufficiently reduced to prevent damages. The fact that corrosion will continue for a certain time has to be taken into account for the design of the repair measure. If corrosion has proceeded so far that limit states like critical cracking will be reached soon, it might be too late to use moisture control, and alternative methods should be used that stop corrosion immediately.

To monitor the rate of corrosion and the influence of moisture control over time, sensors are available that allow us to follow the decrease of the corrosion rate or the increasing resistivity over time (see Section 8.4). The relationship between corrosion risk and electrical resistivity of the concrete can be estimated using Table 6.34 (Langford and Broomfield 1987). However, it has to be considered that this is just a rough estimation, because there are several influencing factors, including geometry of the corrosion system, etc. (Warkus 2013) (Table 6.34).

Table 6.34 Reference values for the relationship between the risk of corrosion-induced damages and the electrical resistivity of the concrete

Electrical resistivity of the concrete	Risk of corrosion-induced damages
Ωm	—
<100	High
100–500	Medium
500–1000	Low
>1000	Negligible

6.10.2 Method 8.1: Hydrophobic impregnation

Hydrophobic impregnation can be used for different principles. While moisture control related to concrete corrosion is Method 2.1, related to reinforcement corrosion it is Method 8.1. For both methods it is important to prevent ingress of water and allow drying out by evaporation through the hydrophobic layer, as shown in Figure 6.70.

Regarding crack treatment, carbonation and appearance advice is given in the section describing Method 1.1. Besides these criteria, it has to be considered whether water may penetrate from the other sides to the concrete element, e.g., by capillary suction from the soil or other sources. If this cannot be avoided, other methods than 8.1 have to be selected to achieve corrosion protection for the reinforcement.

The products for Method 8.1 are standardised in EN 1504-2. Details regarding the performance characteristics of hydrophobic impregnations are given in the sections on Methods 1.1 and 2.1.

This method is especially interesting to protect fair-faced concrete surfaces against carbonation-induced reinforcement corrosion. A side effect of hydrophobic treatments is that the concrete surface is to a certain extent water repellent and self-cleaning. Due to the reduction of the water content with time, the concrete appears more homogenous and light.

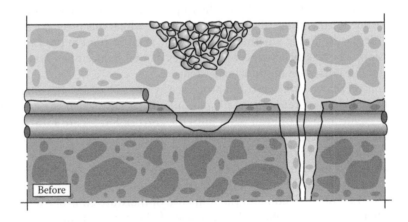

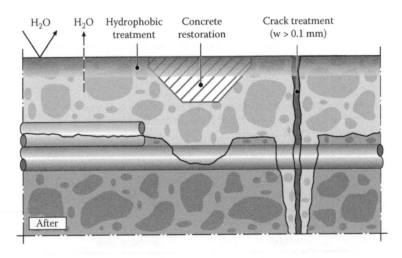

Figure 6.70 Schematic representation of Method 8.1 before and after application.

Table 6.35 Summary of Method 8.1

Method 8.1: Hydrophobic impregnation.
Principle 8: Increasing resistivity.
Approach: Reduction of the corrosion rate of the reinforcement by drying of the concrete and subsequent increasing of the electrical resistivity.
Typical applications: Carbonation-induced reinforcement corrosion at fair-faced concrete surfaces at an early stage.
Special attention should be paid to:
• **Design:** As drying takes time, corrosion will continue after treatment and slow down slowly; reserves in load-bearing capacity are required.
• **Product requirements:** According to EN 1504-2.
• **Execution:** Careful surface preparation; concrete surface must be sufficiently dried out; high penetration depth shall be targeted.
• **Quality control:** Depth of penetration, hydrophobicity, etc.
Durability/maintenance: Regular inspections recommended.
Complementary methods: Usually Methods 1.5 and 3.1 to 3.3.

However, to ensure durable corrosion protection of the reinforcement by increased resistivity, regular inspections and monitoring are recommended. Furthermore, it has to be taken into account that the reinforcement will continue corroding for a certain time, until the resistivity has increased so far that the corrosion rates are negligible. As this can take some months or even a few years, a structural analysis must prove that sufficient reserves in cross section of the reinforcement are available.

To monitor the effectiveness of this method, sensors indicating the electrical resistivity, like multiring electrodes or corrosion sensors indicating the corrosion rate of the reinforcement, can be used (see Sections 3.2.8.6 and 8.4). These sensors will show the process of drying over time and indicate in time when the hydrophobic effect gets lost and the hydrophobic treatment has to be renewed. However, a sufficient amount of sensors must be placed in representative areas to get reliable information on the status of the reinforcement regarding corrosion. Actually, research is running on nondestructive measurement of the hydrophobic effect by mobile NMR technology (Antons et al. 2012) (Table 6.35).

6.10.3 Method 8.2: Impregnation

Similar to moisture control related to concrete corrosion (Method 2.2), an impregnation can also be used to increase the electrical resistivity of the concrete and reduce the corrosion rate of the reinforcement, until a harmless level is reached. To achieve a high degree of pore filling with the impregnation material, the concrete surface has to be prepared as shown in Figure 6.71. Weak concrete areas have to be restored and cracks closed.

It has already been explained in the sections of impregnations for other principles that impregnations are not able to bridge crack movements. As soon as new cracks develop or existing cracks get wider, the impregnated concrete will crack and water can penetrate again. For this reason, the fields of application of Method 8.2 are restricted to situations where new cracks or crack movements can be excluded.

Compared to a hydrophobic treatment, drying out of the concrete is expected to be significantly slower using impregnations because the goal of impregnations is to fill the pores. This will allow only low evaporation rates.

If new cracks or crack movements cannot be ruled out, coatings with adequate crack bridging ability can be used to increase resistivity, according to Method 8.3 (Table 6.36).

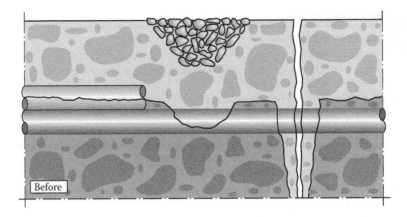

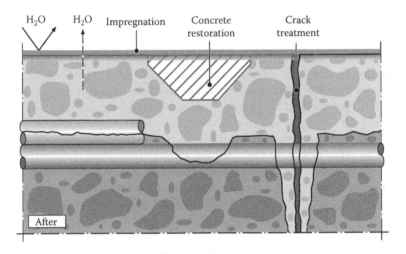

Figure 6.71 Schematic representation of Method 8.2 before and after application.

Table 6.36 Summary of Method 8.2

Method 8.2: Impregnation.

Principle 8: Increasing resistivity.

Approach: Closing the pores of the concrete surface to reduce the water content, increase electrical resistivity of the concrete and decrease the corrosion rate of the reinforcement.

Typical applications: Limited applications; floors, horizontal surfaces.

Special attention should be paid to:

- **Design:** No protection when cracks move or new cracks occur; as drying takes time, corrosion rates will decrease slowly.
- **Product requirements:** According to EN 1504-2.
- **Execution:** Careful surface preparation; concrete surface must be sufficiently dried out.
- **Quality control:** Depth of penetration, film thickness, etc.

Durability/maintenance: High durability, depending on use.

Complementary methods: If required, Methods 1.5 and 3.1 to 3.3.

6.10.4 Method 8.3: Coating

Besides hydrophobic impregnations and pore-filling impregnations, coating systems can be used to protect the reinforcement by increasing resistivity of the concrete. Compared to the two previously described alternatives, coating systems can be used in many fields of applications, because their performance can be adjusted to nearly all conditions in actual practice.

To achieve an increasing electrical resistivity of the concrete by drying out, the coating systems should be impermeable to water ingress and, as much as possible, open for evaporation of water vapour from the concrete. However, high evaporation rates are contradicting to other performance criteria and are not always possible. Experience has shown that coating systems, which are impermeable for water ingress and have a low evaporation rate, also lead to an increasing electrical resistivity, but they require longer times for drying out what is shown later.

The application of this method is shown schematically in Figure 6.72. As usual, before application of the concrete, weak concrete has to be restored and cracks have to be closed.

The materials for coatings are standardised in EN 1504-2. If drying of the concrete shall proceed quickly, coatings with high evaporation rates should be selected.

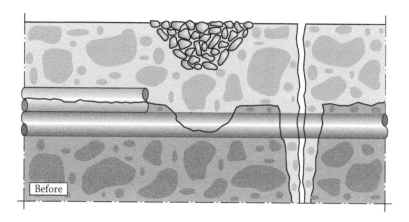

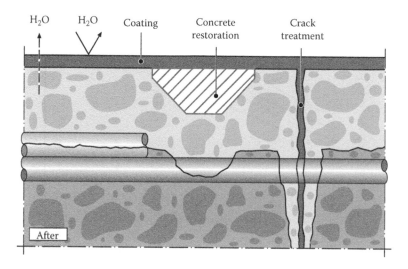

Figure 6.72 Schematic representation of Method 8.3 before and after application.

In the case of chloride-induced corrosion, this method should be treated with great care, because chlorides are hydrophilic and decrease the rates of evaporation significantly. At moderate or even high chloride contents this method cannot be recommended. However, if it is intended to use it regardless of this fact, there is a certain risk that corrosion will continue at reduced rates and cause corrosion damages after some time. To control this situation, extensive inspection and monitoring are required.

In the following example, Method 8.3 has been tested for a parking deck. As a coating system, a crack bridging system consisting of different layers of reactive polymers and a thickness of about 5 mm has been used (see Section 5.5.4). The permeability against water of such a coating system is very low. To follow the process of drying, multiring electrodes have been installed into the concrete below the coating system. These allow a direct measurement of the electrical resistance of the concrete in the course of time. These data can be used to calculate the specific resistivity of the concrete.

Figure 6.73 shows the development of electrical resistances over seven years after application of the coating system. The so-called cell constant of the multiring electrode is about 10 cm, so the resistivity can be calculated by multiplying the measured resistances with a factor of 0.10 m. The distances between the measuring rings A1 and A6 were 5 mm, allowing the measurement of resistance profiles in 5 mm steps.

Figure 6.73 shows that the resistances between the rings increase from less than 1000 Ωm to values that are considerably higher than 1000 Ωm after about 1–3 a. The values between rings A1 and A2 were infinite, which can be explained by the effect that ring A1 probably has been sealed by the coating system.

Comparing these high resistivities with the values given in Table 6.34 leads to the estimation that after about 3 a only negligible corrosion rates have to be expected, and that the corrosion rates are still decreasing after about 7 a. As already mentioned for Methods 8.1 and 8.2, it has to be considered for this method that during the time of continuing corrosion (about 3 a), a certain loss of cross section of the reinforcement occurs. This reduces the load-bearing capacity of the structural element.

For a more precise estimation of the corrosion rates, it is recommended to perform additional electrochemical measurements (Table 6.37).

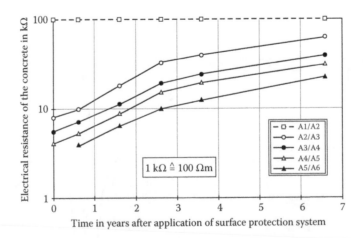

Figure 6.73 Electrical resistances of the concrete of a chloride-contaminated parking structure after application of a coating system, measured with a multiring electrode in different depths over time.

Table 6.37 Summary of Method 8.3

Method 8.3: Coating.

Principle 8: Increasing resistivity.

Approach: Application of a coating system, which prevents water ingress and allows evaporation of water from the concrete to increase electrical resistivity of the concrete and to limit the corrosion rate of the reinforcement.

Typical applications: Standard method for carbonation-induced corrosion; for chloride-induced corrosion only applicable in an early stage with low chloride contents.

Special attention should be paid to:

- **Design:** Requirements have to be specified (EN 1504-2).
- **Product requirements:** According to EN 1504-2.
- **Execution:** Careful surface preparation; concrete surface must have the required wetness; minimum thickness must be ensured.
- **Quality control:** Adhesion to concrete, coating thickness, etc.

Durability/maintenance: Inspections are recommended, depending on use.

Complementary methods: Usually Methods 1.5 and 3.1 to 3.3.

6.11 PRINCIPLE 9: CATHODIC CONTROL

6.11.1 General

The informative annex A of EN 1504-9 gives some explanations of this principle: The approach of Principle 9 is creating conditions in which potentially cathodic areas of reinforcement are unable to drive an anodic reaction. It relies upon restricting access of oxygen to all potentially cathodic areas, to the point when corrosion cells are stifled and corrosion is prevented by the inactivity of the cathodes.

Saturation of the whole of a self-contained reinforced concrete unit is an example of the application of this principle. Limiting oxygen content by saturating the concrete should be used only where the whole of the member is under water and where reinforcement of the submerged member is electrically isolated from all reinforcement in members that are not submerged, or where there is no effective return path for ionic currents through the concrete.

Even in this immersed situation, the risk of corrosion may be increased by the presence of contaminants such as chloride ions, and additional repair principles may be required.

While surface coatings can be applied to concrete structures that have a low oxygen diffusion rate, in practice it may be difficult to control corrosion by this method, particularly where chloride ions are present. The quality of surface preparation and application needs to be to a very high standard to provide a defect-free barrier.

It has to be noted that application of a coating that is a partial barrier to oxygen may restrict the selection of other repair principles; an impressed current cathodic protection system, e.g., is unlikely to work if applied over an oxygen resisting coating.

These explanations of EN 1504-9 show that the fields of application for this principle are rare, and that great care has to be taken if it is used. However, there may be some fields of applications at structures in permanently wet soils or under water.

6.11.2 Method 9.1: Limiting oxygen content (at the cathode) by saturation or surface coating

Figure 6.74 shows schematically the application of Method 9.1. If the concrete surface is permanently in contact with water, the concrete will be water saturated and the corrosion

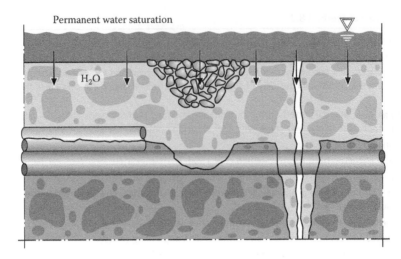

Permanent water saturation

H$_2$O

Figure 6.74 Schematic representation of Method 9.1.

Table 6.38 Summary of Method 9.1

Method 9.1: Limiting oxygen content (at the cathode) by saturation or surface coating.

Principle 9: Cathodic control.

Approach: Reducing the corrosion rate by preventing the access of oxygen at the surface of the reinforcement by saturation or coating.

Typical applications: No reports on applications available.

Special attention should be paid to:

- **Design:** Macrocells and oxygen supply must be excluded for the whole remaining service life.
- **Product requirements:** No special requirements.
- **Execution:** No special requirements.
- **Quality control:** No special requirements.

Durability/maintenance: Regular inspections and monitoring recommended.

Complementary methods: Additional cathodic protection may not be possible.

rates of the reinforcement will be uncritical. As described above, it has to be ensured that no significant amounts of oxygen reach the reinforcement from the back side or other sources, and that the formation of macrocells can be excluded. The use of a coating to limit the oxygen content is not shown in Figure 6.74 because no reports on systematic research on this method or on experience from practice are available. Therefore, the use of a coating for this method has to be treated carefully.

To check whether macrocells are active, potential measurements or other electrochemical methods like galvanic current measurements can be used. However, the evaluation of such readings requires special knowledge in the field of corrosion of the reinforcement (Table 6.38).

6.12 PRINCIPLE 10: CATHODIC PROTECTION

6.12.1 General

The informative annex A of EN 1504-9 gives some explanations of this principle: The approach of Principle 10 is creating conditions in which potentially anodic areas of reinforcement are unable to take part in the corrosion reaction. It is especially appropriate where chloride contamination is significant. Cathodic protection (CP) can control corrosion regardless of the level of chloride contamination in the concrete and limits the amount of chloride removal to that physically damaged by corrosion of underlying reinforcement. Its long-term effectiveness depends on adequate monitoring and maintenance.

Cathodic protection is standardised in the international performance standard EN ISO 12696, which covers all aspects of CP in reinforced concrete structures from structure assessment and repair, CP system components, installation procedures, commissioning, system records and documentation, to operation and maintenance.

For Principle 10, only Method 10.1 is available, which is called "applying an electrical potential." However, in the following the method is simply called cathodic protection (CP).

6.12.2 Method 10.1: Applying an electrical potential

For cathodic protection an anode has to be installed close to the reinforcement that shall be protected. This can be achieved by inserting anodes into drilled holes or slots or by mounting anodes on top of the concrete surface, shown schematically in Figure 6.75. As an

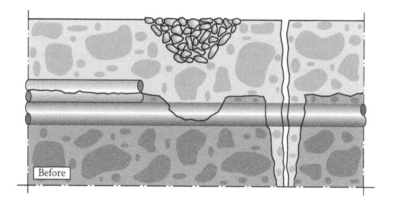

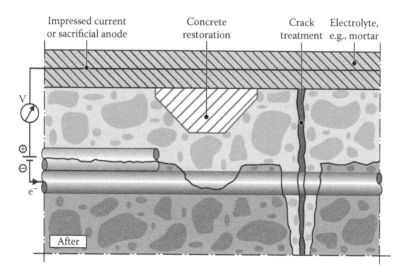

Figure 6.75 Schematic representation of Method 10.1 using a metal anode embedded in mortar before and after application.

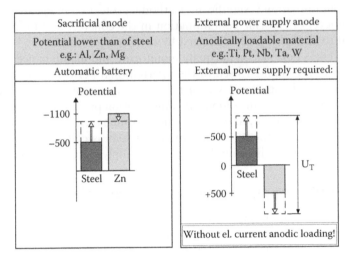

Figure 6.76 Schematic representation of cathodic protection using sacrificial anodes or an impressed current system.

example for an anode system, a metal anode embedded in mortar is shown to demonstrate the principle. In this case concrete with insufficient strength has to be restored and cracks have to be closed to allow the application of the mortar layer.

Generally cathodic protection can be achieved by the use of sacrificial anodes or impressed current systems, as shown in Figure 6.76. Materials for sacrifical anodes like aluminium, zinc, or magnesium have a more negative rest potential than corroding reinforcing steel. If the sacrificial anode is connected with the reinforcement, a galvanic cell is formed like a battery, and the sacrificial anode will start to corrode and protect the reinforcement cathodically. One advantage of a sacrificial anode system is that no external current supply is required, but only an electrical connection between sacrificial anode and reinforcement. On the other side, the lifetime of the sacrificial anode is limited because it is consumed slowly by corrosion.

Impressed current systems are working differently. An external voltage is applied between anodes and reinforcement. As anodes, materials that are stable under anodic polarisation, like titanium, platinum, niobium, tantalum, or wolfram, are used. The potential of the reinforcement is polarised cathodically, as shown in Figure 6.76. It should be noted that a connection of the electrochemically noble anodes with the reinforcement without external voltage would polarise the reinforcement anodically, resulting in an acceleration of corrosion.

Impressed current systems have the advantages that they are not consumed by corrosion, provided that they are suitable for CP, and that the protective electrical current can be controlled by selecting the external voltage. Therefore, today most CP systems for steel in concrete are impressed current systems. The electrical currents required for CP are very low, and the power consumption is usually lower than 100 W/1000 m^2 of the concrete surface to be protected. If no power supply is available, the required electricity can be provided in most cases by solar panels.

During operation of the cathodic protection (CP) system several electrochemical processes occur, which are shown schematically in Figure 6.77.

These are generally the same processes as described in Section 6.9.4 dealing with Method 7.3. Generally, it can be distinguished between primary and secondary protection effects (Bruns and Raupach 2009). Primary protection effects are:

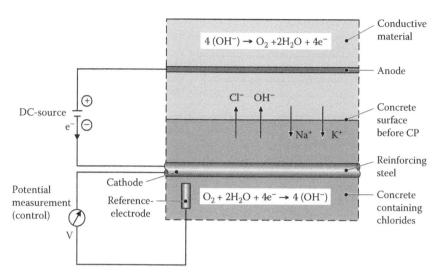

Figure 6.77 Electrochemical processes occurring during operation of a CP system.

- Cathodic polarisation of the depassivated reinforcement directly hinders the anodic dissolution of the steel by moving the reaction equilibrium on the steel surface toward the cathodic reaction (oxygen reduction).
- Cathodic polarisation of the passive reinforcement surfaces leads to an equilisation of the potentials of active and passive steel surfaces. The potential difference between those areas that is the driving force for macroelement corrosion is significantly reduced.

Secondary effects are:

- The formation of hydroxide ions in the cathodic reaction leads to an increase of the pH value in the pore solution at the steel surface, which has a stabilising effect on the passive film.
- Due to the electric field, anions, like chloride ions, migrate from the reinforcement toward the anode, and cations, like sodium and potassium ions, migrate toward the reinforcement. Over the long term this leads to a decrease of the chloride content at the steel surface.
- Due to the increased consumption of oxygen by the cathodic reaction under specific circumstances (wet concrete, high concrete cover), the oxygen concentration at the steel surface may decrease significantly, which makes cathodic polarisation of the reinforcement easier.

Experience from CP systems over more than 20 years shows that the secondary effects lead to a significant improvement of the conditions regarding corrosion. Actually, it is discussed that CP systems may be switched off after some years of operation.

During recent years several types of anode systems have been developed, as shown schematically in Figure 6.78.

Generally CP anodes should be homogenously distributed close to the reinforcement to be protected. However, to prevent short circuits there must be a certain minimum distance to the reinforcement.

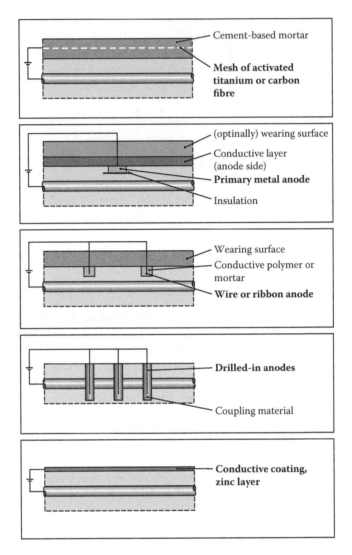

Figure 6.78 Schematic representation of typical types of anode systems for CP.

Mesh anodes made of activated titanium embedded in mortar are the most widely used anode systems. They are very long-term stable and provide a homogenous current distribution. Figure 6.79 shows an example for such a mesh anode on a ramp of a parking garage. The mesh is fixed to the concrete surface and connected to a titanium ribbon, as seen in the middle of Figure 6.79.

This mesh anode is embedded in mortar to achieve the required electrolytic conductivity for CP. The requirements for the activated titanium and the mortar are given in EN ISO 12696. Additionally, for the mortar the requirements according to EN 1504-3 as repair mortar have to be fulfilled (Methods 3.1 to 3.3, depending on the type of application).

Finally, a coating system is applied on the mortar that prevents further ingress of chlorides according to Method 1.3 and improves the appearance.

Cables are led from the reinforcement, the sections of the anodes, the reference cells, and the monitoring sensors to a central location where the power supply is installed. Today remote-controlled systems are available, allowing us to check the monitoring data and

Figure 6.79 Mesh anode made of activated titanium on a ramp of a parking garage before application of the mortar.

adjust the voltages of each protection zone individually from the office. The possibility to monitor CP systems allows a permanent evaluation of the effectiveness of the system. This is an important advantage of CP compared to other methods without monitoring.

Alternatively to the traditional anode meshes, ribbon anodes are also used, as shown in Figure 6.80. All ribbons within one protective zone have to be connected electrically to be able to transfer the protective electrical current. The two connection ribbons can be seen in Figure 6.80. The distances of the ribbons have to be small enough to ensure a homogenous current distribution within the reinforcement. Like the anode mesh, the ribbons are embedded in suitable mortar.

Besides mounting anodes to the concrete surface, we can also install them in slots. However, this requires sufficient concrete cover to ensure that the reinforcement is not damaged and no electrical short circuits occur. Figure 6.81 shows a bridge with slots already prepared to insert ribbon anodes. Before slotting, the concrete cover has been checked over the full surface.

Figure 6.80 Ribbon anode made of activated titanium on a parking deck before application of the mortar. (From Bruns, M., in *10. Symposium Kathodischer Korrosionsschutz von Stahlbetonbauwerken*, ed. S. Gieler-Breßmer, Esslingen, November 22–23, 2012, pp. 55–63, Ostfildern: Techniche Akademie Esslingen, 2012.)

Figure 6.81 Slotted anodes made of activated titanium on a bridge deck before application of the sealing. (From Bruns, M., in *3. Kolloquium Erhaltung von Bauwerken: Conservations of Buildings*, ed. M. Raupach, Esslingen, January 22–23, 2013, pp. 69–75, Ostfildern: Technische Akademie Esslingen, 2013.)

The ribbon anodes are inserted into the slots and connected to ribbon anodes running perpendicular to the slots shown in Figure 6.81, which act as current distributors. By short-circuit measurements it can permanently be checked that the ribbon anodes have no electrical connection to the reinforcement. Afterwards, the slots are sealed with a conductive material.

In cases where no sufficient concrete cover is available to use slotted anodes, and where it is difficult to apply mortar to the concrete surface, drilled-in core anodes can be used. Figure 6.82 shows a hollow box girder, which is difficult to access due to limited space and

Figure 6.82 Drilled-in core anodes made of activated titanium in a box girder of a bridge with difficult access. (From Bruns, M., Binder, G., *Beton- und Stahlbetonbau* 108 (2013), no. 2, pp. 104–115.)

high-voltage cables running through. Applying mortar and preparing the surface of the concrete are practically impossible under these conditions. To protect the reinforcement against corrosion due to chlorides that have penetrated into the hollow box girder through leaking joints, core anodes have been installed into drilled holes in the bottom slab and the lower parts of the walls. In Figure 6.82 the cables to the anodes indicate the sensor positions.

Using this CP system with core anodes, it was possible to protect the reinforcement safely under the given difficult working conditions.

To design the CP system, potential mapping has been carried out and the positions of the reinforcement have been determined using a cover meter. A numerical simulation has been carried out to design the geometry of the anodes and the optimal positions of the anodes.

Cathodic protection can also be used as a preventive measure before corrosion has started. In such cases the protective electrical current is much smaller than when the reinforcement is already corroding.

A special CP application is shown in Figure 6.83. Concrete damaged by alkali-silica reaction had to be replaced by new concrete at a pillar in seawater. To prevent the formation of a macrocell between the reinforcement in the remaining concrete below the repair area and the reinforcement of the new concrete above, ribbon anodes have been placed at the bottom of the new concrete (see Figure 6.84). After application of an impressed current between these anodes and the reinforcement, a potential barrier is created that prevents the new concrete from acting as macro-cathode accelerating corrosion in the old concrete.

The monitoring data of the CP system (potentials, currents, etc.) show that the system is working as designed.

Besides the anode systems described above, also anodes based on carbon meshes, carbon-filled paints, zinc, and other materials are available. The examples show that cathodic protection systems can be used in various situations to protect the reinforcement against chloride-induced corrosion (Table 6.39).

Figure 6.83 Pillar damaged by alkali-silica reaction before and after removing the concrete by high-pressure water jetting. (From Bruns, M., Raupach, M., *Restoration of Buildings and Monuments* 15 (2009), no. 5, pp. 355–366.)

Figure 6.84 Preventive CP with ribbon anodes made of activated titanium in the pillar before placement of the concrete. (From Bruns, M., Raupach, M., *Restoration of Buildings and Monuments* 15 (2009), no. 5, pp. 355–366.)

Table 6.39 Summary of Method 10.1

Method 10.1: Applying an electrical potential.

Principle 10: Cathodic protection.

Approach: Cathodic polarisation of the reinforcement until the corrosion rate gets uncritical.

Typical applications: All types of concrete structures exposed to chlorides.

Special attention should be paid to:
- **Design:** According to EN ISO 12696.
- **Product requirements:** According to EN ISO 12696.
- **Execution:** According to EN ISO 12696.
- **Quality control:** According to EN ISO 12696.

Durability/maintenance: According to EN ISO 12696.

Complementary methods: Usually Methods 1.5 and 3.1 to 3.3.

6.13 PRINCIPLE 11: CONTROL OF ANODIC AREAS

6.13.1 General

The informative annex A of EN 1504-9 gives some explanations of this principle: Where contamination of the concrete is extensive, but it is not possible to remove all contaminated concrete, incipient anode formation can be controlled by treating the surface of the reinforcement in the patch repair to prevent corrosion. Coatings can be applied directly to the reinforcement where it is exposed as part of concrete restoration. These coatings can obtain active pigments, which may function as anodic inhibitors or by sacrificial galvanic action.

Other types of coating can form barriers on the surface of the concrete. This method can only be effective if the reinforcement is prepared to be free of corrosion and the coating is complete; i.e., the bar must be completely encapsulated and the coating is defect-free. The method should not be considered unless the entire circumference of the reinforcing bar can be coated. The effect of the coating on bond between the reinforcement and concrete should also be considered.

Alternatively, corrosion inhibitors can be used that chemically change the surface of the steel or form a passive film over it. Corrosion inhibitors can be introduced either by addition to the concrete repair product or system or by application to the concrete surface followed by migration to the depth of the reinforcement. Inhibitors that are applied to the surface of the concrete must penetrate the concrete down to the level of the reinforcement to take effect. There is currently no standard for inhibitors, so evidence of the effectiveness of any such product should be obtained before specifying its use. Note that some corrosion inhibitors work by control of both anodic and cathodic areas. In severe conditions, additional repair principles may be required.

In the following sections the three previously mentioned methods for Principle 11 are presented and discussed.

6.13.2 Method 11.1: Active coating of the reinforcement

Method 11.1 is schematically shown in Figure 6.85. It involves several working steps: At first the reinforcement needs to be uncovered so far that sufficient space is available around the rebar to apply an active coating. Before coating, the surface of the reinforcement has to be cleaned carefully from rust and loose particles.

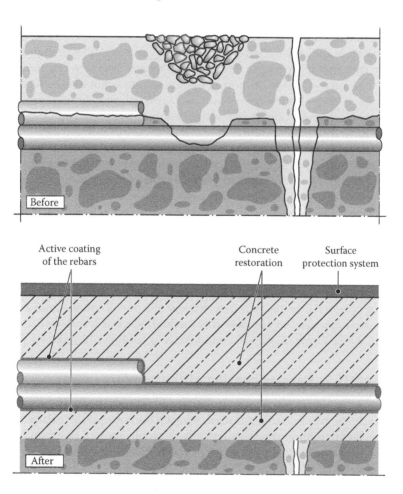

Figure 6.85 Schematic representation of Method 11.1 before and after application.

Table 6.40 Summary of Method 11.1

Method 11.1: Active coating of the reinforcement.
Principle 11: Control of anodic areas.
Approach: Corrosion protection of the reinforcement by an active coating.
Typical applications: When not sufficient, concrete cover is available or can be provided; temporary protection of exposed reinforcement.
Special attention should be paid to:
  • **Design:** Unknown durability, etc.
  • **Product requirements:** According to EN 1504-7.
  • **Execution:** Careful cleaning of the reinforcement; careful application.
  • **Quality control:** Thickness of the active coating, etc.
Durability/maintenance: Regular inspections recommended.
Complementary methods: Additionally, Method 1.3 recommended.

As active coatings, different types can be used. Widely used are cement-based coatings with a high pH value that can promote repassivation of the reinforcement. Furthermore, zinc-based materials are available that actively protect the reinforcement against corrosion by a certain level of cathodic protection. Finally, coatings containing anodic inhibitors may be used. The active coatings are standardised in EN 1504-7. However, the materials should be selected carefully.

After application of the active coating the concrete cover needs to be restored using suitable mortar or concrete according to Methods 3.1 to 3.3 (concrete restoration). It has to be considered that usual concrete or mortar for restoration is not able to bridge significant crack movements. If critical crack movements are expected, further measures are required. Finally, often a surface protection system is applied to prevent open cracks and to improve the appearance as well as the durability.

Generally corrosion protection of the reinforcement can be provided by the mortar or concrete for restoration alone, without an active coating (Method 7.2). However, if the concrete cover is too small to preserve passivity over the remaining service life, an additional coating can be applied on top of the concrete (Principle 1.3) or an active coating could be applied on the reinforcement (Method 11.1). To ensure durability in practice, often both methods (1.3 and 11.1) are applied.

Another field of application for Method 11.1 is when the reinforcement is uncovered for a certain time until the concrete is applied to close the openings. In this case corrosion of the reinforcement during the exposed phase can be prevented by the active coating.

The durability of the active coating alone, without taking the protective effects of the concrete and the optional coating on top of the concrete into account, is not known exactly. Therefore, it is recommended to combine this method, when it is not only used for temporary protection, with coating on the concrete surface according to Method 1.3 (Table 6.40).

6.13.3 Method 11.2: Barrier coating of the reinforcement

Barrier coatings are in theory electrically isolating and prevent the anodic dissolution of iron as well as the cathodic reaction (reduction of oxygen). Such properties can be achieved, e.g., by epoxy resins. Figure 6.86 shows schematically the application of this method. Analogous to Method 11.1 the reinforcement has to be exposed, cleaned, and coated. Afterwards the opening has to be closed with a suitable mortar or concrete.

While this principle works well in theory and under perfect working conditions, in practice it is difficult to achieve a sufficient quality of the barrier coating, especially on the back

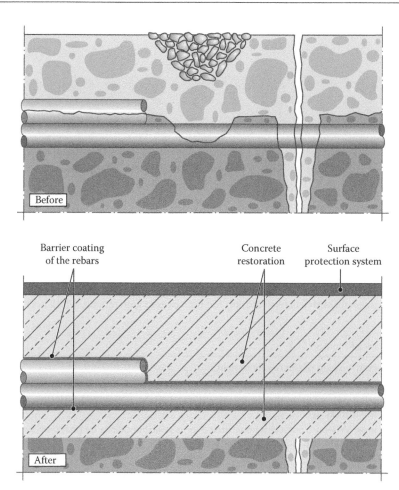

Figure 6.86 Schematic representation of Method 11.2 before and after application.

side of crossings or areas with many reinforcing bars in a close distance. If the quality of the barrier coating is insufficient, e.g., if the thickness is too small or adhesion is bad, there is a high risk of corrosion of the reinforcement under the coating. In this case the alkalinity of the repair concrete cannot passivate the reinforcement, because it is shielded by the coating. This can lead to serious corrosion of the reinforcement, which can only be detected in a progressed state.

Therefore, it cannot be recommended to use this method when the working conditions are difficult and a sufficient quality of the barrier coating is not guaranteed. However, if Method 11.2 shall be used regardless of this risk, it is recommended to control the quality of the coating works on site very strictly.

Figure 6.87 shows an example where an epoxy-based coating has been applied to the surface of the reinforcement, but without applying mortar to the coated rebars. It can clearly be seen that the coated reinforcement is already corroding. To achieve an adequate corrosion protection on the steel surface, coating systems used also for the protection of steel constructions could theoretically be used. However, most of these systems consist of different layers, starting on the steel surface with a first coating containing active pigments, which would be according to Method 11.1 (Table 6.41).

Figure 6.87 Reinforcement with a barrier coating, but not covered with mortar, which is already corroding.

Table 6.41 Summary of Method 11.2

Method 11.2: Barrier coating of the reinforcement.

Principle 11: Control of anodic areas.

Approach: Corrosion protection of the reinforcement by a barrier coating, which isolates the reinforcement from adverse agents.

Typical applications: When no sufficient concrete cover is available or can be provided; temporary protection of reinforcement exposed for a longer time.

Special attention should be paid to:

- **Design:** Risk of corrosion underneath the coating, when the quality is not sufficient, etc.
- **Product requirements:** According to EN 1504-7.
- **Execution:** Careful cleaning of the reinforcement; careful application.
- **Quality control:** Complete coating with required thickness, adhesion, etc.

Durability/maintenance: Regular inspections recommended.

Complementary methods: As an alternative, Method 11.1 should be considered; additionally, Method 1.3 is recommended.

6.13.4 Method 11.3: Applying corrosion inhibitors in or to the concrete

Method 11.3 covers two ways of the application of corrosion inhibitors in reinforced concrete, as schematically shown in Figures 6.88 and 6.89. They can be applied to the concrete surface (Method 11.3-1) or mixed into repair mortars (Method 11.3-2).

Surface-applied corrosion inhibitors must penetrate through the concrete cover to the reinforcement until the protective amount of inhibitors is reached around the reinforcement. However, especially for concretes with low permeability and high cover thickness, this is a difficult task. Therefore, it is recommended to prepare trial areas and to determine the effectiveness of the corrosion inhibitor by electrochemical corrosion rate measurements.

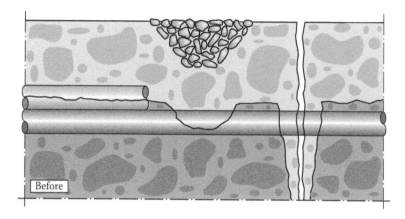

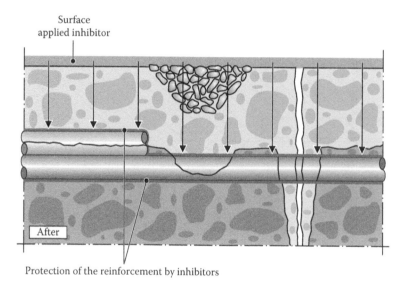

Figure 6.88 Schematic representation of Method 11.3-1 before and after application.

Inhibitors may be mixed with repair mortars or concrete to improve passivity, resulting in an increased level for the critical chloride content, or to reduce possible corrosion rates by increasing the anodic or cathodic polarisation resistance of the reinforcement in concrete. Generally different types of inhibitors are available, as shown in Table 6.42.

Inhibitors are not standardised in EN 1504. Therefore, it is recommended to check the effectiveness and optimise the application parameters within a trial area. If the corrosion rate is only reduced by a certain percentage, it has to be estimated whether this is sufficient for the remaining service life or whether other or additional methods have to be applied.

To prevent leaching out of the inhibitors, usually a coating is applied to the concrete surface after the treatment (Method 1.3) (Table 6.43).

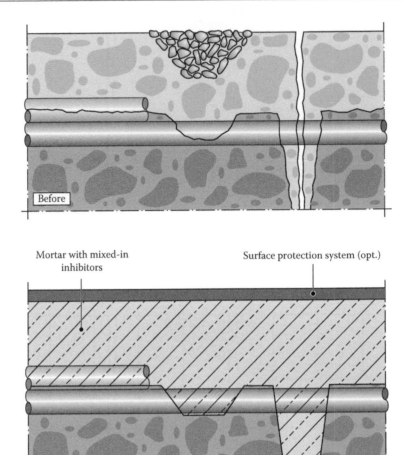

Figure 6.89 Schematic representation of Method 11.3-2 before and after application.

Table 6.42 Types of Corrosion Inhibitors

Type	Mechanism of Action	Example
Physical inhibitors	Blocking or shielding of the metal surface by adsorption	Inorganic or organic ions
Passivators	Thin, dense protective layer (about 20 nm)	E.g., nitrite, chromate
Surface layer creators	Less dense voluminous layers	E.g., phosphate
Electrochemical inhibitors	Thin porous film made of foreign metals by metal exchange reaction	E.g., Hg, As, Sb
De-stimulators	Reaction with the corrosive agents	E.g., O_2 or Cl elimination

Source: Fischer, H., *Werkstoffe und Korrosion* 6 (1955), no. 1, pp. 26–32, 1955.

Table 6.43 Summary of Method 11.3

Method 11.3: Applying corrosion inhibitors in or to the concrete.

Principle 11: Control of anodic areas.

Approach: Prevent corrosion or reduce corrosion rate by applying corrosion inhibitors over the concrete surface or mixing to repair mortars or concrete.

Typical applications: Additional protection to other methods.

Special attention should be paid to:

- **Design:** Effectiveness and durability should be tested for the given conditions (concrete quality, corrosion status, environmental conditions).
- **Product requirements:** Not standardised.
- **Execution:** According to product specifications.
- **Quality control:** Trial areas, etc.

Durability/maintenance: Regular inspections of monitoring recommended.

Complementary methods: Method 1.3 (protection against leaching out).

Chapter 7

Execution and Quality Control

7.1 GENERAL

The execution as well as the quality control of concrete repair is regulated in Part 10 of the EN 1504 norm series. This part features an overview of recommended tests in order to ensure an execution of the work according to current technical standards.

It has to be noted that the general requirements for the quality control are given in EN 1504-10 as follows:

> Consideration shall be given to the chemical, electrochemical and physical condition of the substrate and any contaminants, the ability of the structure to accept loading, movement and vibration during protection and repair, ambient conditions, and the characteristic of the materials contained in the structure and those of the protection and repair products and systems:
>
> The following requirements shall be met:
>
> - The achievement of the required condition of the substrate regarding cleanliness, roughness, microcracking, cracking, tensile and compressive strength, chloride or other contaminant and their penetration, depth of carbonation, moisture content, temperature and degree of corrosion of reinforcement.
> - The achievement of the compatibility of the original concrete and reinforcement with the protection or repair products and systems and compatibility between any different products and systems, including avoiding the risk of creating conditions which may cause corrosion.
> - The achievement of the specified properties of products and systems when applied and in their hardened condition regarding the fulfillment of their purpose for protection and repair of the structure.
> - The achievement of the required storage and application conditions regarding ambient temperature, humidity and dew point, wind force and precipitation and any temporary protection which is needed.

The following sections comprehend the required and recommended investigations and tests to ensure a proper quality. The lists of quality control procedures in (DIN) EN 1504-10:2003-10 feature several options for the need of each procedure (compare to Section 5.2):

- ■ Has to be conducted for all intended uses
- ◆ Has to be conducted if specified
- ◻ Has to be conducted for special uses

7.2 Hydrophobic treatment and impregnation

In order to ensure the quality of hydrophobic treatments (Methods 1.1, 2.1, and 8.1) as well as impregnations (Methods 1.2, 2.2, 5.2, 6.2, and 8.2), the following tests have to be conducted before, during, and after the application (Tables 7.1 to 7.3).

Depending on the boundary conditions, it can be necessary to investigate the crack width and crack movement before applying a hydrophobic treatment. Also, it is recommended to take drill cores in cracked areas in order to detect preliminary crack treatments. During the application it is further recommended to check the appearance of the hydrophobic treatment regularly, and thus detect irregularities within the product as well as the total amount of consumption.

Table 7.1 Properties of the surface before and during the application

Property	Test method	European standard	Need[a]	See section
Surface debonding	Sounding with a hammer	—	■	—
Cleanness	Visual inspection, remove dust by wiping	—	■	—
Surface strength	Determination of the surface strength	EN 1542	♦	3.2.2.4
Moisture content of the surface	Drying method, determination of the electrical resistivity	—	♦	3.2.8
Temperature of the surface	Thermometer	—	■	—
Depth of carbonation	Phenolphthalein test	EN 14630	♦	3.2.7.2
Chloride content	Extraction of concrete samples	EN 14629	♦	3.2.7.3

[a] See Section 7.1.

Table 7.2 Boundary conditions before and during application

Property	Test method	European standard	Need[a]	See section
Temperature	Thermometer	—	■	—
Relative humidity	Hygrometer	—	■	—
Rain	Visual inspection	—	■	—
Wind velocity	Anemometer	—	■	—
Dew point	Thermometer, hygrometer	—	♦	7.3.2

[a] See Section 7.1.

Table 7.3 Properties of the hydrophobic treatment after application

Property	Test method	European standard	Need[a]	See section
Ingress depth of the impregnation agent	Extraction of drill core, total amount of consumption	EN 12504-1, EN ISO 2808	♦	7.3.3
Water permeability	Karsten's tube, water uptake	EN 12390-1, ISO 7031	■	3.2.8.5

[a] See Section 7.1.

7.3 QUALITY CONTROL PARAMETERS BEFORE AND AFTER APPLICATION OF COATINGS

7.3.1 Roughness

The roughness of the surface has to be determined according to (DIN) EN 1766:2000 by using the so-called sand spot method. The method uses the following formula to calculate the roughness by the volume of the sand as well as the diameter of an even sand spot made on the surface.

$$R_t = \frac{40 \cdot V}{\pi \cdot d^2}$$

where

R_t = Roughness of the surface in mm
V = Volume of sand in mm³
d = Diameter of the sand die in mm

The volume of the sand (usually quartz sand with a grain size between 0.05 and 0.1 mm) is determined by using a scale or a measuring cup. Typical volumes of sand used are between 5 and 25 ml. The sand is poured on the concrete surface, and by using a wooden die an even spot is generated. Then the diameter can be determined by using a scale, and the roughness can be calculated as well.

7.3.2 Dew point temperature

The dew point temperature is one essential parameter regarding the execution of hydrophobic treatments, impregnations, as well as coatings. In general, the repair or protection product cannot be applied if the ambient dry temperature is less than 3 K above the dew point temperature. The dew point temperature can be determined in dependency of the ambient dry temperature as well as the relative humidity (rh) according to Table 7.4.

According to (DIN) EN 1504-10:2003 the air temperature has to be measured using a mercury or digital thermometer. The thermometer has to have a required accuracy of ±0.5°C. The surface temperature may be measured with a digital electronic thermometer with a required accuracy of ±0.5°C. The relative humidity should be measured according to ISO 4677 Parts 1 and 2.

Table 7.4 Overview of the dew point temperature in dependency of the ambient dry temperature as well as the relative humidity according to (DIN) EN 1504-10:2003

Ambient dry temperature	Dew point temperature (°C) for ambient relative humidity between 40 and 100% rh						
	40%	50%	60%	70%	80%	90%	100%
35	19.4	23.0	26.1	28.7	31.0	33.1	35
30	15.0	18.5	21.4	23.9	26.2	28.2	30
25	10.5	13.9	16.7	19.6	20.1	23.2	25
20	6.0	9.3	12.0	14.4	16.5	18.3	20
15	1.5	4.2	7.3	9.6	11.6	13.4	15
10	−3.0	0.1	2.6	4.8	6.7	8.5	10
5	−7.0	−4.7	−2.0	0	1.9	3.5	5

7.3.3 Thickness of the freshly applied and hardened coating

The thickness of a coating is one key parameter for the properties of a surface protection system. The thickness can be measured during as well as after the application of the coating. All test methods can be found in (DIN) EN ISO 2808:2007. The following sections describe selected and most commonly used methods.

The thickness of a coating during the application can be approximated either by the total consumption or by using a wet film comb gauge as shown in Figure 7.1. The gauge is placed upright in the freshly applied coating, and thus the fresh coating will wet all teeth that are closer to the surface than the thickness of the coating itself. The tooth that is least wetted indicates the thickness of the fresh coating. Wet film comb gauges are available in different

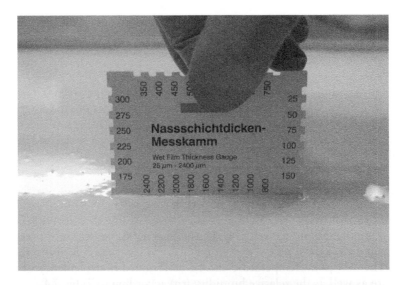

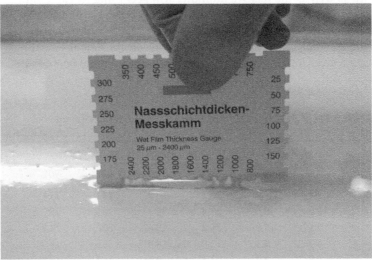

Figure 7.1 Measuring the thickness of a freshly applied surface protection system with a comb gauge. Top: Completely immersed gauge. Bottom: Gauge removed out of the surface protection system.

scales, usually beginning at a minimum thickness of 10 μm and ending at a maximum of 2500 μm. Instead of comb gauges, wheel type gauges are also available.

The thickness of the hardened coating can be determined on site by performing an angled cut and measuring the length of the visible cut face by using a magnifying glass or by peeling. As long as the angle of the cut is known, the thickness can be calculated by Pythagoras's law. Also, a small hole in the coating can be made and the thickness of the coating can be measured by a micrometre depth gauge.

Alternatively, a small area of the coating can be peeled off the surface and the thickness can be determined with a micrometre gauge. Figure 7.2 illustrates a peeled-off coating. Besides the thickness of the coating, the adhesion can be judged qualitatively. In the presented case, the thickness of the coating did not fulfil the requirements and the adhesion was also very low and not in compliance with the requirements.

Another possibility is to extract small drill cores out of the structure and determine the thickness of the coating in the lab by using a calliper gauge or even a microscope. For

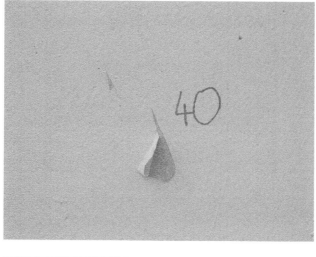

Figure 7.2 Determination of the adhesion (qualitatively) as well as the thickness of the coating on site. Top: Cut before peeling off manually. Bottom: During peeling off the surface protection system.

Figure 7.3 Determination of the impregnation depth of a hydrophobic treatment. Top: Overview. Bottom: Detailed view.

hydrophobic treatments the extraction of a drill core is the only reliable method to determine the impregnation depth. In order to do so, the core is cracked into two halves and water is sprayed on the surfaces. The hydrophobic areas will remain lighter due to the water repellency, and the nonhydrophobic areas will turn dark due to the water uptake (see Figure 7.3).

If a film-forming surface protection system has to be investigated, the thickness can easily be determined on the cut face of the drill core (see Figure 7.4).

7.3.4 Appearance of the coating

The appearance of a film-forming coating has to be determined according to the norm series EN ISO 4628, which distinguishes the following defects of the appearance:

- Degree of blistering (EN ISO 4628-2)
- Degree of rusting (EN ISO 4628-3)
- Degree of cracking (EN ISO 4628-4)
- Degree of flaking (EN ISO 4628-5)
- Degree of chalking by tape method (EN ISO 4628-6)
- Degree of chalking by velvet method (EN ISO 4628-7)
- Degree of delamination and corrosion around a scribe or another artificial defect (EN ISO 4628-8)

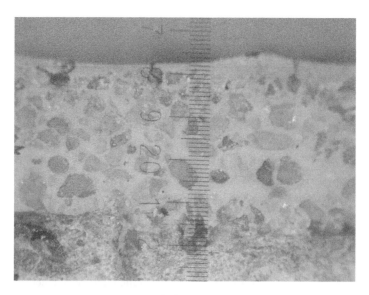

Figure 7.4 Determination of the thickness of a film-forming surface protection system on cut surfaces. Top: Drivable surface protection system without crack bridging ability (OS-8). Bottom: Surface protection system for parking decks (crack bridging class B3.2 (−20°C) according to (DIN) EN 1062-7:2004).

The judgement of each type of defect is done based on reference pictures given in each part of the norm series. These pictures can also be used as a reference for an automated image analysis. The quantification itself is done with a standardised system according to Part 1 of the norm series EN ISO 4628.

This standardised system differs between the amount of defects, the size of defects, and relative changes between two different times of observation. Tables 7.5 to 7.7 give an overview of the corresponding specific values.

In the report the type of defect is specified, and additionally the classification according to the previously mentioned system is always given, e.g.:

Table 7.5 Specific values for determining the amount of defects according to (DIN) EN ISO 4628-1:2003

Specific value	Amount of defects
0	No defects visible
1	Very little amount of defects
2	Little, but significant, amount of defects
3	Moderate amount of defects
4	Considerable amount of defects
5	Large number of defects

Table 7.6 Specific values for determining the size of defects according to (DIN) EN ISO 4628-1:2003

Specific value	Size of defects (if not stated differently in another part of EN ISO 4628)
S0	Not visible at a magnification ratio of 10 times
S1	Only visible up to a magnification ratio of 10 times
S2	Barely visible with the human eye
S3	Clearly visible with the human eye
S4	Size between 0.5 and 5 mm
S5	Size larger than 5 mm

Table 7.7 Specific values for determining the intensity of change according to (DIN) EN ISO 4628-1:2003

Specific value	Intensity of change
0	No visible change of appearance
1	Very little change, barely visible
2	Little change, clearly visible
3	Medium change, very clearly visible
4	Strong and distinct changes
5	Very strong changes

- Blistering 2 (S2) → little, but significant amount of blistering, which is barely visible with the human eye
- Flaking 3 (S4) → moderate amount of flaking with a size between 0.5 and 5 mm

Figures 7.5 to 7.11 show typical defects according to EN ISO 4628 of film-forming surface protection systems.

7.3.5 Adhesion of the coating

The adhesion of a film-forming surface protection system can be determined either quantitatively by a uniaxial tensile test (see Section 3.2.2.4) or qualitatively by the crosscut test according to (DIN) EN ISO 2409:1992.

The crosscut test quantifies the adhesion of a coating after the coating is cut in a grid consisting of 12 perpendicular lines (master plate; see Figures 7.12 and 7.13) and a tape is glued to the surface and ripped off the cut surface. The classification is then done according to Table 7.8, which defines a maximum amount of flaking for each class.

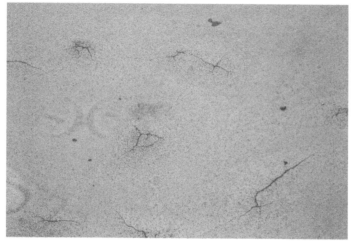

Figure 7.5 Examples of blistering on the surface of a surface protection system.

Figure 7.6 Example of rusting.

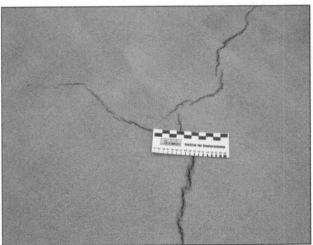

Figure 7.7 Example of cracks in surface protection systems.

7.4 COATING

In order to ensure the quality of coatings (Methods 1.3, 2.2, 5.1, 6.1, 7.1, 8.2, and 9.1), the following tests have to be conducted before, during, and after the application (Tables 7.9 to 7.11).

Depending on the boundary conditions, it can be necessary to investigate the crack width. Also, it is recommended to take drill cores in cracked areas in order to detect preliminary crack treatments. During the application it is further recommended to check the appearance of the coating treatment regularly, and thus detect irregularities within the coating as well as the total amount of consumption.

Figure 7.8 Example of flaking.

Figure 7.9 Example of chalking.

7.5 FILLING CRACKS, VOIDS, OR INTERSTICES

In order to ensure the quality of filled cracks, voids, or interstices (Methods 1.5, 4.5, and 4.6), the following tests have to be conducted before, during, and after the application (Tables 7.12 to 7.14).

Before the application, it is recommended to take drill cores in cracked areas in order to detect preliminary crack treatments. Also, during the application it is recommended to check the appearance of the coating treatment regularly, and thus detect irregularities within the product as well as the total amount of consumption.

Figure 7.10 Examples of delamination of a surface protection system (Wolf and Raupach, 2008).

7.6 APPLICATION OF MORTAR OR CONCRETE

In order to ensure the quality of mortar or concrete used for concrete repair (Methods 3.1, 3.2, 3.3, 4.4, 5.1, 6.1, 7.1, 7.2, and 7.4), the following tests have to be conducted before, during, and after the application (Tables 7.15 to 7.17).

Depending on the boundary conditions, it can be necessary to determine the humidity of the concrete as well as the crack width and possible crack treatments before applying mortar or concrete. After the application, besides the mentioned properties, the layer thickness should also be measured.

7.7 ADDING OR REPLACING REINFORCING BARS

In order to ensure the quality of added or replaced reinforcing bars (Method 4.1), the following tests have to be conducted before, during, and after the application (Tables 7.18 to 7.20).

Figure 7.11 Example of distinct change of colour due to sunlight exposure (Wolf and Raupach, 2008).

7.8 ADDING REINFORCEMENT ANCHORED IN HOLES

In order to ensure the quality of added reinforcement anchored in holes (Method 4.2), the following tests have to be conducted before, during, and after the application (Tables 7.21 to 23).

7.9 BONDING PLATE REINFORCEMENT

In order to ensure the quality of bonding plate reinforcement (Method 4.3), the following tests have to be conducted before, during, and after the application (Tables 7.24 to 7.26).

7.10 COATING OF THE REINFORCEMENT

If reinforcement is coated (Methods 11.1 and 11.2), the quality of the coating is ensured by the following tests having to be conducted before, during, and after the application (Tables 7.27 to 7.29).

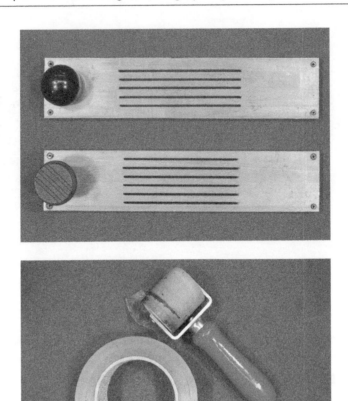

Figure 7.12 Top: Master plates required for the crosscut test. Bottom: Tape and roller required for the test after cutting.

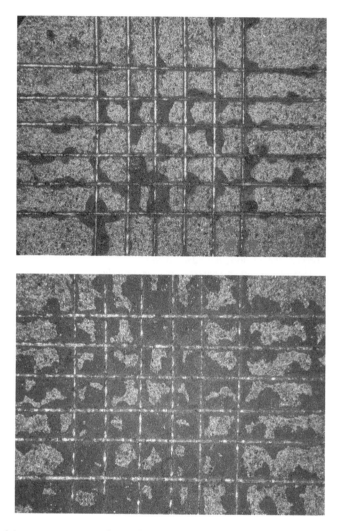

Figure 7.13 Result of the crosscut test performed on coated steel. Top: classification 2. Bottom: classification 4.

Table 7.8 Classification of the crosscut test results according to ISO 2409:1992

Classification	Description	Appearance of surface of crosscut area from which flaking has occurred
0	The edges of the cuts are completely smooth; none of the squares of the lattice are detached.	
1	Detachment of small flakes of the coating at the intersection of the cuts. A crosscut area not significantly greater than 5% is affected.	
2	The coating has flaked along the edges at the intersections of the cuts. A crosscut area significantly greater than 5%, but not significantly greater than 15% is affected.	
3	The coating has flaked along the edges of the cuts partly or wholly in large ribbons, or it has flaked partly or wholly on different parts of the squares. A crosscut area significantly greater than 15%, but not significantly greater than 35%, is affected.	
4	The coating has flaked along the edges of the cuts in large ribbons, or some squares have detached partly or wholly. A crosscut area significantly greater than 35%, but not significantly greater than 65%, is affected.	
5	Any degree of flaking that cannot be classified in classification 4.	—

Table 7.9 Properties of the surface before and during the application

Property	Test method	European standard	Need[a]	See section
Surface debonding	Sounding with a hammer	—	■	—
Cleanness	Visual inspection, remove dust by wiping	—	■	—
Surface evenness	Visual inspection	—	■	—
Roughness	Visual inspection, sand test	EN 1766	◆	7.3.1
Surface strength	Determination of the surface strength	EN 1542	◆	3.2.2.4
Crack movements	Mechanical gauges or linear variable differential transformer (LVDTs)	—	□	3.2.5.5
Moisture content of the surface	Drying method, determination of the electrical resistivity	—	◆	3.2.8
Temperature of the surface	Thermometer	—	■	—
Ingress of harmful substances	Extraction of concrete samples	—	◆	—

[a] See Section 7.1.

Table 7.10 Boundary conditions before and during application

Property	Test method	European standard	Need[a]	See section
Temperature	Thermometer	—	■	—
Relative humidity	Hygrometer	—	◆	—
Rain	Visual inspection	—	■	—
Wind velocity	Anemometer	—	■	—
Dew point	Thermometer, hygrometer	—	◆	7.3.2
Thickness of the freshly applied coating	Measuring comb	EN ISO 2808	◆	7.3.3

[a] See Section 7.1.

Table 7.11 Properties of the coating after application

Property	Test method	European standard	Need[a]	See section
Thickness of the hardened coating	V-notch test, total amount of consumption	EN ISO 2808	■	7.3.3
Appearance of the coating	Visual inspection	EN ISO 4628-1, …, EN ISO 4628-6	■	7.3.4
Water permeability	Karsten's tube, water uptake	EN 12390-8, ISO 7031	◆	3.2.8.5
Adhesion of the coating	Tape test, adhesion test	EN ISO 2409, EN ISO 4624, EN 1542	■	7.3.5

[a]See Section 7.1.

Table 7.12 Properties of the surface before and during the application

Property	Test method	European standard	Need[a]	See section
Cleanness	Visual inspection	—	◆	—
Crack width and depth	Ultrasonic measurements, extraction of drill cores, visual inspection	EN 12504-1, prEN 12504-4, ISO 8047	◆	3.2.5.3
				3.2.5.4
Crack movements	Mechanical gauges or LVDTs	—	◆	—
Moisture content of cracks and surrounding concrete	Drying method, determination of the electrical resistivity	—	◆	—
Temperature of the surface	Thermometer	—	◆	
Pollution of cracks	Extraction of drill cores, chemical analysis	—	◆	

a   See Section 7.1.

Table 7.13 Boundary conditions before and during application

Property	Test method	European standard	Need[a]	See section
Temperature	Thermometer	—	■	—
Relative humidity	Hygrometer	—	◆	—
Rain	Visual inspection	—	◆	—

a   See Section 7.1.

Table 7.14 Properties of the crack filling after application

Property	Test method	European standard	Need[a]	See section
Water permeability	Karsten's tube, water uptake	EN 12390-8, ISO 7031	◆	3.2.8.5
Filling level	Extraction of drill cores, ultrasonic measurement	EN 12504-1, prEN 12504-4:1998, ISO 8047	◆	3.2.2.2
Adhesion of filling material on the surface	Inspection of drill cores and visual inspection	EN 12504	☐	—

a   See Section 7.1.

Table 7.15 Properties of the surface before and during the application

Property	Test method	European standard	Need[a]	See section
Surface debonding	Sounding with a hammer	—	■	—
Cleanness	Visual inspection	—	■	—
Roughness	Visual inspection, sand test	EN 1766	◆	7.3.1
Surface strength	Determination of the surface strength	EN 1542	◆	3.2.2.4
Crack movements	Mechanical gauges or LVDTs	—	□	3.2.5.5
Vibrations of the construction	Acceleration gauge	—	□	—
Temperature of the surface	Thermometer	—	■	—
Depth of carbonation	Phenolphthalein test	EN 14630	□	3.2.7.2
Chloride content	Extraction of concrete samples	EN 14629	□	3.2.7.3
Ingress of harmful substances	Extraction of concrete samples	—	□	—
Electrical resistivity	Wenner test method	—	□	3.2.7.5
Compressive strength	Extraction of drill cores, rebound hammer	EN 12504-1, EN 12504-2	◆	3.2.2.3

[a] See Section 7.1.

Table 7.16 Boundary conditions before and during application

Property	Test method	European standard	Need[a]	See section
Temperature	Thermometer		■	—
Rain	Visual inspection		■	—
Concrete viscosity	Slump, flow test	EN 12350-1, −5, EN 13395-3	■	—
Mortar viscosity	Slump, flow test	EN 13395-1, −2, −4	■	—
Air content	Pressure pot	EN 12350-7	◆	—
Concrete thickness and concrete cover	Extraction of drill cores, visual inspection, magnetic methods	EN 12504-1	■	3.2.3
Compressive strength	Cubic specimens, rebound hammer	EN 12390-1, −2, −3, EN 12190, EN 12504-2	■	3.2.2.3

[a] See Section 7.1.

Table 7.17 Properties of the mortar/concrete after application

Property	Test method	European standard	Need[a]	See section
Surface debonding	Sounding with a hammer	—	■	—
Electrical resistivity	Wenner test method	—	□	3.2.7.5
Water permeability	Karsten's tube, water uptake	EN 12390-8, ISO 7031	◆	3.2.8.5
Concrete cover	Extraction of drill cores, visual inspection, magnetic methods	EN 12504-1	■	3.2.3
Adhesion of filling material on the surface	Inspection of drill cores and visual inspection	EN 12504	■	3.2.2.4
Compressive strength	Drill cores, rebound hammer	EN 12390-1, −2, −3, EN 12190, EN 12504-2	■	3.2.2.3
Gross density	Drying method	EN 12390-7	■	—
Cracks induced by shrinkage	Visual inspection	—	■	3.2.5.3
Voids behind repair patches	Ultrasonic measurements, ground-penetrating radar (GPR), extraction of drill cores	prEN 12504-4, ISO 8047, EN 12504-1	◆	3.2.2.2
Colour and texture of the surface	Visual inspection	—	◆	—

a See Section 7.1.

Table 7.18 Properties of the surface before application

Property	Test method	European standard	Need[a]	See section
Cleanness of the existing rebars	Visual inspection	EN ISO 8501-1	■	—
Dimension of the existing rebars	Visual inspection	—	■	—
Corrosion status of the existing rebars	Half-cell potential, visual inspection	—	◆	3.2.7.4 3.2.7.6

a See Section 7.1.

Table 7.19 Boundary conditions before and during application

Property	Test method	European standard	Need[a]	See section
Rain	Visual inspection	—	■	—
Position of the reinforcement	Extraction of drill cores, visual inspection, magnetic methods	—	■	3.2.3

a See Section 7.1.

Table 7.20 Properties of the replaced rebars after application

Property	Test method	European standard	Need[a]	See section
Position of the reinforcement	Extraction of drill cores, visual inspection, magnetic methods	—	■	3.2.3
Adhesion to the concrete	Pullout test	prEN 1881-1	◆	—

a See Section 7.1.

Table 7.21 Properties of the surface before and during the application

Property	Test method	European standard	Need[a]	See section
Cleanness	Visual inspection	—	■	—
Roughness	Visual inspection, sand test	EN 1766	♦	7.3.1
Moisture content of the surface	Drying method, determination of the electrical resistivity	—	♦	3.2.8 3.2.7.5
Dimension of the existing rebars	Visual inspection	—	■	—
Corrosion status of the existing rebars	Half-cell potential, visual inspection	—	♦	3.2.7.4 3.2.7.6
Cleanness of the rebars	Visual inspection	ENV ISO 8502-1, EN ISO 8502-2, −3, −4	■	—

a See Section 7.1.

Table 7.22 Boundary conditions before and during application

Property	Test method	European standard	Need[a]	See section
Temperature	Thermometer	—	■	—
Relative humidity	Hygrometer	—	■	—
Rain	Visual inspection	—	♦	—
Viscosity of the mortar used for anchoring the rebars	Slump, flow test	EN 12350-1, −5, EN 13395-3	■	—
Position of the reinforcement	Extraction of drill cores, visual inspection, magnetic methods	—	♦	3.2.3

a See Section 7.1.

Table 7.23 Properties of the reinforcement anchored in holes after application

Property	Test method	European standard	Need[a]	See section
Position of the reinforcement	Extraction of drill cores, visual inspection, magnetic methods	—	■	3.2.3
Adhesion to the concrete	Pullout test	prEN 1881-1	♦	—

a See Section 7.1.

Table 7.24 Properties of the surface before and during the application

Property	Test method	European standard	Need[a]	See section
Surface debonding	Sounding with a hammer	—	■	—
Cleanness of the surface and the bonding plates	Visual inspection, remove dust by wiping	—	■	—
Surface evenness	Visual inspection	—	■	—
Roughness	Visual inspection, sand test	EN 1766	■	7.3.1
Surface strength	Determination of the surface strength	EN 1542	■	3.2.2.4
Crack movements	Mechanical gauges or LVDTs	—	◆	3.2.5.5
Vibrations of the construction	Acceleration gauge	—	◆	—
Depth of carbonation	Phenolphthalein test	EN 14630	◆	3.2.7.2
Moisture content of the surface	Drying method, determination of the electrical resistivity	—	■	3.2.8 3.2.7.5
Temperature of the surface	Thermometer	—	■	—
Chloride content	Extraction of concrete samples	EN 14629	◆	3.2.7.3
Corrosion status of the existing rebars	Half-cell potential, visual inspection	—	◆	3.2.7.4 3.2.7.6
Compressive strength	Extraction of drill cores, rebound hammer	EN 12504-1, EN 12504-2	◆	3.2.2.3

[a] See Section 7.1.

Table 7.25 Boundary conditions before and during application

Property	Test method	European standard	Need[a]	See section
Temperature	Thermometer	—	■	—
Relative humidity	Hygrometer	—	■	—
Rain	Visual inspection	—	■	—
Dew point	Thermometer, hygrometer	—	■	7.3.2

[a] See Section 7.1.

Table 7.26 Properties of the bonding plate reinforcement after application

Property	Test method	European standard	Need[a]	See section
Thickness of the coating applied to the bonding plates	V-notch test, total amount of consumption	EN ISO 2808	◆	7.3.3
Voids between surface and bonding plate	Impact-echo test, sounding with a hammer, ultrasonic measurement	prEN 12504-24:1998, ISO 8047	■	3.2.2.2
Load-bearing capacity	Load test	—	◆	—

[a] See Section 7.1.

Table 7.27 Properties of the surface before and during the application

Property	Test method	European standard	Need[a]	See section
Cleanness	Visual inspection, remove dust by wiping	—	■	—
Vibrations of the construction	Acceleration gauge	—	◆	—
Temperature of the surface	Thermometer	—	■	—

[a] See Section 7.1.

Table 7.28 Boundary conditions before and during application

Property	Test method	European standard	Need[a]	See section
Temperature	Thermometer	—	■	—
Relative humidity	Hygrometer	—	■	—
Rain	Visual inspection	—	◆	—
Dew point	Thermometer, hygrometer	—	◆	7.3.2

[a] See Section 7.1.

Table 7.29 Properties of the coating after application

Property	Test method	European standard	Need[a]	See section
Thickness of the hardened coating	V-notch test, total amount of consumption	EN ISO 2808	◆	7.3.3
Appearance of the coating	Visual inspection	EN ISO 4628-1, ..., EN ISO 4628-6	■	7.3.4

[a] See Section 7.1.

Chapter 8

Maintenance of Concrete Structures

8.1 GENERAL

In order to achieve the desired lifetime after a repair of a structure, it is necessary to define a maintenance plan. The maintenance plan is thus an essential element of the repair of a structure and should not be neglected. EN 1504 defines the key issues of a maintenance plan in Part 9, which describes the necessity of maintenance as follows:

> Some parts of the protected or repaired concrete structure may have an expected service life which is short compared with that of the rest of the concrete structure. Familiar examples are surface coatings, sealants, and weather proofing materials. If the integrity of the protection or repair depends on such parts, it is essential that they be regularly inspected, tested and renewed if necessary.

It is also states that "a structure management strategy is not chosen on technical grounds alone, but also on economic, functional, environmental and other factors, and most importantly the owner's requirements for the structure."

8.2 KEY ISSUES OF A MAINTENANCE PLAN

According to Part 9 of EN 1504, the maintenance plan should include the following aspects. The following listing gives information for future maintenance which should be included:

a) an estimate of the expected remaining design life of the concrete structure;
b) identification of each part whose design life is expected to be less than the required service life of the concrete structure;
c) the date at which each such part is next to be inspected or tested;
d) the system of inspection which is to be used, how results are to be recorded and how future inspection dates are to be decided;
e) a specification for continuous treatment (if any is required), for example cathodic protection;
f) a statement of precautions to be taken or prohibitions to be enforced, for example maintenance of surface water drainage, maximum pressure for washing or prohibition of the use of de-icing salt.

A maintenance management system should also be implemented to ensure that the required maintenance is carried out properly, because a "correct monitoring and maintenance of the protection and repair works will result in a longer service life for both the works and the structure."

8.3 MAINTENANCE CONCEPTS

The maintenance concepts of a structure can be designed and conducted in different ways which all have the same goal: to reach, the agreed limit states at the end of the intended lifetime. Thus, it is necessary to define the desired strategy directly after the repair in order to be able to judge the time-dependent development of the structure's condition by using the previously described inspection and diagnoses technologies. These technologies are very sophisticated, so a definition of the status of the entire structure or parts of the structure can be done with high accuracy.

In order to predict the time-dependent evolution of the status of the structure, various models are available. A very general time-dependent evolution of the structure condition is given in Section 9 of EN 1504. For example, Frangopol (2000) contains more specific time-dependent curves, which will be discussed in the following paragraphs. It is important that the selected model does predict the time-dependent evolution of the main cause of deterioration or the combination of causes that lead to a degradation of the structure. This makes the selection of the proper model even more complex because mostly the structure is not damaged by a single degradation cause, such as ingress of chlorides.

It is also very important to define the limit state properly. Generally, the limit state is defined not only by the condition of the load-bearing elements, but also by requirements given by the appearance as well as the utilisation of the structure, which might even change over the lifetime. If this happens, it might be necessary to adapt the definition of the limit state and adjust the maintenance concept.

Figure 8.1 shows three different types of strategy (Frangopol 2000):

1. Reactive maintenance strategy
2. Preventive maintenance strategy
3. Combination of both

The reactive maintenance strategy features the repair only at the point when the actual status of the structures reaches the limit state, and not before. After reaching the desired level of structural status at the time being, the process begins again. The limit state is usually defined by the occurrence of visible signs of concrete corrosion, e.g., cracks or spalling, or due to rebar corrosion, e.g., signs of rust on the surface.

The preventive maintenance plan features the repair of the structure without any visible damages, e.g., renewal of a surface protection system after 5 or 10 years even though the surface protection system might still fulfil its requirements. The advantage of this strategy is that the construction might not suffer great damage, and thus the lifetime of the entire structure can be enlarged significantly. Also, it is possible to plan the maintenance work in advance and reduce costs due to unscheduled closings of the structure, e.g., parking decks.

The combination of both strategies can also be feasible if, e.g., the lifetime of the construction is reduced after the initial repair, and thus the effort after the first preventive maintenance due to returning work should be minimised. But as also shown in Figure 8.1, depending on the time-dependent conditional of the structure, the amount of maintenance work must not be limited.

One key element of a maintenance concept is a continuous monitoring of the conditions of the structure. The following section will present selected types of monitoring systems.

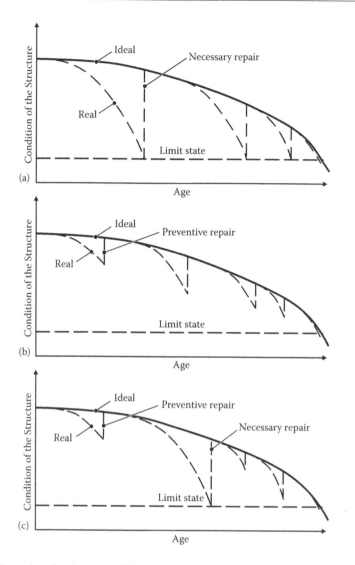

Figure 8.1 Time-dependent development of the status of a structure due to different maintenance concepts. (From Frangopol, D.M., in *Proceedings of the 6th International Workshop on Material Properties and Science: Present and Future of Health Monitoring*, ed. P. Schwesinger, F.H. Wittmann, Weimar, September 2000, pp. 9–20, Freiburg: ADIFICATIO, 2000.)

8.4 IMPLEMENTATION OF SENSOR-BASED MONITORING SYSTEMS

8.4.1 General

A regular assessment of the status of a structure can be done with the methods described in Section 3. Especially in areas that are difficult to access, sensors that determine certain material or structural properties continuously are feasible and enable a precise knowledge of the condition of the structure. Also, the installation of sensors enables a continuous recording of relevant parameters without any man-made influences. It has to be noted that a regular inspection, as well as sensors, does not eliminate any accuracy during the repair process. Both are integral parts of the entire concept of maintaining the health of a structure over the desired life span.

The market features an endless number of different systems, which cannot be named entirely, so that the following sections only give a broad overview of the most commonly used monitoring systems.

8.4.2 Monitoring of the structural behaviour

The structural behaviour is usually quantified by length changes, elongations of structural elements, or changes of the static or dynamic load-bearing behaviour. The changes of length of a beam or any other type of structural element can be measured by using LVDTs or strain gauges. LVDTs are generally applied on the surface of the structure (see Section 3, Figure 3.43).

Strain sensors can be either applied on the surface of the construction or alternatively embedded in the structure (see Figure 8.2). Both sensors measure the strain by changes of the electrical resistivity of the sensor itself. The electrical resistivity changes proportional to the cross section, which changes due to elongation or compression of the structure.

As previously described, the elongation of a structural element can be measured by strain gauges or linear variable differential transformers (LVDTs), but these sensors only measure in a very defined area and not the elongation of an entire beam, for example. In order to measure the elongation of such large structural elements entirely, fibre-optic systems are used. Fibre optics can be placed in the structural element during production or analogue to LVDTs along the surface. The elongation is then measured by using different technologies (see detailed information, e.g., Matthys et al. (2006). All technologies are based on the fact that light is passed through the fibre optics, and due to deformations of the structure, the fibre optics, and thus the light, changes its properties. Possible properties are:

- Intensity of light transmitted through the fibres
- Phase of the light

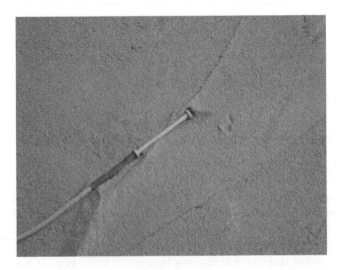

Figure 8.2 Strain sensor embedded between two layers of PCC—sensor length 50 mm.

- Refractive index of the fibre by using fibre Bragg grating (FBG)
- Amount of light scattered and reflected back to the light source, so-called optical time-domain reflection (OTDR) sensors

The selection of the proper type of sensor mainly depends on the desired results, as well as the boundary conditions of the task itself. The load-bearing behaviour of a structure can be measured indirectly by the deformations of the structure, but also directly by using load cells or acceleration sensors. Acceleration sensors can be either piezoelectric or acoustic sensors. Especially the influence of acoustic sensors by external noise has to be regarded during designing the sensor setup.

8.4.3 Monitoring of relevant concrete parameters

Besides the load-bearing capacity, the following concrete parameters are also relevant concerning the durability of the structure:

- Water content
- pH value
- Chloride content

The water content of concrete can be measured by the previously described multiring electrodes, which are based on measuring the concrete resistivity and calculating the water content based on calibration curves. Other sensors that are based on the same principle are listed in, e.g., McCarter et al. (2001).

It is also possible to install regular moisture sensors in boreholes and measure the relative humidity in the boreholes. Basically, this measuring technique can be calibrated, but investigations reveal that the changes of humidity in the borehole are not identical to the changes of the water content of the material.

If leakages in extensive sealants should be detected, electrical sensors have to be installed along the entire sealant. The pH value as well as the chloride content of the concrete can be measured indirectly with reference electrodes (see Section 6.12).

8.4.4 Monitoring of the corrosion behaviour of the reinforcement

Modern sensor systems allow us to monitor the risk of corrosion of the steel reinforcement, and thus to initiate proper protective measures before noteworthy damage occurs (see Figure 2.12). The results of the monitoring can be used to verify the time-dependent development of the status of a structure (see Figure 8.1) in terms of design verification. This allows a constant update of the models that are used to calculate the time-dependent evolution of the structural condition by a comparison between the predicted and measured properties of the structure. It also allows us to update the model by the results of the monitoring, and thus to enhance the precision of the model. This process is called model update.

The two most important properties, which should be monitored after repairing a concrete structure damaged by rebar corrosion, are the potential of the rebar and the corrosion speed. In order to monitor the time-dependent development of the potential, so-called reference electrodes can be installed.

In order to monitor the corrosion speed of the reinforcement, sensors can be used based on continuous corrosion-current measurements. In order to do so, an artificial macro-element is created that consists of an anode (existing rebar) as well as various cathodes (the sensor itself). The amount of sensors has to be determined by the boundary conditions of the structure.

In order to monitor whether the rebar remains passive after a repair, the cathode made of metal nobler than the reinforcement is connected with the reinforcement and embedded in the structure. Then the current between the anode and cathode is measured. If the current remains high, it can be assumed that the reinforcement is depassivated and the success of the repair has to be questioned. Thus, a strategy to achieve a passive reinforcement has to be developed. If the current is reduced to a negligible value, it can be assumed that the reinforcement in the repaired area is passive or corroding with uncritical speed.

Figure 8.3 shows such a sensor. The sensor itself is usually embedded in mortar (approximately 1 cm) in order to reduce the risk of shortages as well as damages of the cathode during installation. Also, the electrolytic connection between the sensor and the concrete can be ensured by the mortar applied under laboratory conditions.

After placing the sensors in boreholes, the sensors (metal and mortar) are fully embedded by an anchoring mortar and connected to an electronic containing box (so-called terminal box), which enables the connection of a measuring device. The terminal box allows the determination of the currents between the reinforcement and the sensor itself. The determination can be done continuously or in regular intervals, e.g., every half year. Figure 8.4 illustrates the schematic setup of such a sensor network containing the sensors, the terminal box, the connection to the reinforcement, as well as the measuring device.

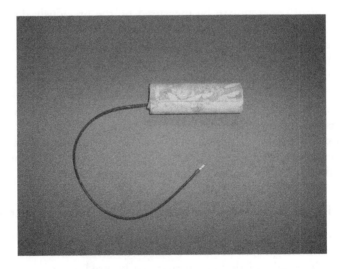

Figure 8.3 Sensor for monitoring corrosion rates of the reinforcement.

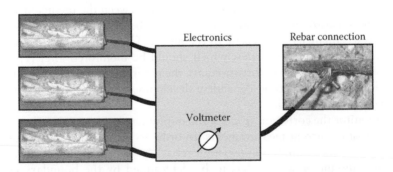

Figure 8.4 Schematic drawing of the entire setup. (Raupach, Gulikers and Reidiling 2013)

Literature, standards, and guidelines

Angst, U., Elsener, B., Larsen, C.K., Vennesland, O. Critical Chloride Content in Reinforced Concrete—A Review. *Cement and Concrete Research* 39 (2009), no. 12, pp. 1122–1138.

Antons, U., Orlowsky, J., Raupach, M. A Non-Destructive Test Method for the Performance of Hydrophobic Treatments. In *Proceedings of the 3rd International Conference on Concrete Repair, Rehabilitation and Retrofitting (ICCRRR)*, Cape Town, South Africa, September 3–5, 2012. London: CRC Press, Taylor & Francis Group, 2012.

Archie, G.E. The Electrical Resistivity Log as an Aid in Determining Some Reservoir Characteristics. *Transactions of the American Institute of Mining and Metallurgical Engineers* 146 (1942), s. 54–62.

Bepple, S. CFK-Lamellen in der Praxis—Anwendung und baubetrieblicher Ablauf. In *Betoninstandsetung heute und für die Zukunft. 11. Fachsymposium am*, Dortmund, March 25, 2003.

Bornstedt, H. Zerstörungsfreies Auffinden von korrodierender Bewehrung bei chloridbeaufschlagten Stahlbetonbauteilen mit Hilfe des Potentialmessverfahrens. Master's thesis, RWTH Aachen University, Institute of Building Materials Research, 1988.

Breit, W. Untersuchungen zum kritischen korrosionsauslösenden Chloridgehalt für Stahl in Beton. In *Schriftenreihe Aachener Beiträge zur Bauforschung*. Dissertation, Institut für Bauforschung der RWTH Aachen, 1997, no. 8.

Bruns, M. Erster Einsatz des KKS in Deutschland an einer vorgespannten Brückenfahrbahnplatte. In *10. Symposium Kathodischer Korrosionsschutz von Stahlbetonbauwerken*, ed. S. Gieler-Breßmer, Esslingen, November 22–23, 2012, pp. 55–63. Ostfildern: Techniche Akademie Esslingen, 2012.

Bruns, M. Einsatz des Kathodischen Korrosionsschutzes bei der Spannbeton-Fahrbahnplatte einer Kanalbrücke. In *3. Kolloquium Erhaltung von Bauwerken: Conservations of Buildings*, ed. M. Raupach, Esslingen, January 22–23, 2013, pp. 69–75. Ostfildern: Technische Akademie Esslingen, 2013.

Bruns, M., Binder, G. Umsetzung des Kathodischen Korrosionsschutzes an den Spannbetonüberbauten der Schleusenbrücke Iffezheim. *Beton- und Stahlbetonbau* 108 (2013), no. 2, s. 104–115.

Bruns, M., Raupach, M. Cathodic Protection at the Eider-Flood Barrage as Barrier against Macrocell Corrosion. *Restoration of Buildings and Monuments* 15 (2009), no. 5, pp. 355–366.

Bruns, M., Raupach, M., Grünzig, H., Schneck, U. Realkalisieren von Fassadenflächen—Ergebnisse eines Forschungsvorhabens. *Restoration of Buildings and Monuments: An International Journal* 11 (2005), no. 5, pp. 285–295.

Catharin, P., Federspiel, H. Der elektrische Widerstand des Betons. *Elektrotechnik und Maschinenbau* 89 (1972), no. 10, s. 399–407.

CEB-FIB Model Code 2010. CEB Bulletin 66.

RL-S1B 2001 DAfStb-Guideline. *Protection and Repair of Concrete Structures*. Berlin, 2001.

RL-S1B 2013 DAfStb-Guideline. *Protection and Repair of Concrete Structures*. Draft 2013.

Deutscher Beton-Verein (DBV), DBV Merkblattsammlung. *Merkblatt Rissbildung: Begrenzung der Rißbildung im Stahlbeton- und Spannbetonbau*. Wiesbaden: Deutscher Beton- und Bautechnik-Verein e.V., 2006.

European Federation of Corrosion Working Party 11. Corrosion of Steel in Concrete. *Materials and Corrosion* 64 (2013), no. 2.

Fischer, H. Inhibition und Inhibitoren. *Werkstoffe und Korrosion* 6 (1955), no. 1, pp. 26–32.

Frangopol, D.M. Bridge Health Monitoring and Life Prediction Based on Reliability and Economy. In *Proceedings of the 6th International Workshop on Material Properties and Science: Present and Future of Health Monitoring*, ed. P. Schwesinger, F.H. Wittmann, Weimar, September 2000, pp. 9–20. Freiburg: ADIFICATIO, 2000.

Glass, G.K., Reddy, B. The Influence of the Steel Concrete Interface on the Risk of Chloride Induced Corrosion Initiation. In *Corrosion of Steel in Reinforced Concrete Structures*, COST 521, Final Workshop, ed. R. Weydert, Luxembourg, February 18–19, 2002, pp. 227–232. Luxembourg: University of Applied Sciences, 2002.

Haardt, P., Hilsdorf, H.K. Realkalisierung karbonatisierter Betonrandzonen durch großflächigen Auftrag zementgebundener Reparaturschichten. In *Werkstoffwissenschaften und Bausanierung, Tagungsbericht des dritten Internationalen Kolloquiums*, ed. F.H. Wittmann, W.J. Bartz, part 1, pp. 666–684. Ehningen: Expert, 1993.

Harnisch, J., Raupach, M. Investigations into the Time to Corrosion and Corrosion Initiating Chloride Contents for Steel in Concrete. In *From the Earth Depths to Space Heights, EUROCORR 2010*, Moscow, September 13–17, 2010. Moscow: Congress Center of World Trade Center, 2010.

Harnisch, J., Raupach, M. Investigations into the Time Dependent Surface Morphology of Reinforcing Steel Subjected to Chloride Containing Concrete. In *Developing Solutions for the Global Challenge, EUROCORR 2011*, Stockholm, September 4–8, 2011. Stockholm: Swerea KIMAB, 2011.

Laase, H., Stichel, W. Stützwandsanierung in Berlin: Spezielle Aspekte der Korrosion an der Rückwand. In *Bautechnik* 60 (1983), Nr. 4, S. 124–129, 1983.

Langford, P., Broomfield, J.P. Monitoring the Corrosion of Reinforcing Steel. *Construction Repair* 1 (1987), no. 2, pp. 32–36.

Kosalla, M., Raupach, M. Probabilistic Life-Cycle Design—Recent Developments in Durability Considerations Regarding Chloride-Induced Corrosion. In *Safer World through Better Corrosion Control, EUROCORR 2012*, Istanbul, September 9–13, 2012. Istanbul: EFC, 2012.

Matthys, S., Somers, R., Taerwe, L., Polen, J.-J. *Optical Fibre Strain Sensing for Monitoring of Concrete Structures*, pp. 279–295. Stuttgart: Fraunhofer-Informationszentrum Raum und Bau, 2006.

McCarter, W.J., Chrisp, T.M., Starrs, G., Basheer, P.A.M., Blewett, J. Smart Structures: A Sensor System to Monitor Covercrete Condition. In *Non-Destructive Testing of Concrete, Fifth CANMET/ACI International Conference on Recent Advances in Concrete Technology*, Singapore, ed. P.A.M Basheer, July 29–August 1, 2001, pp. 31– 51. Farmington Hills, MI: American Concrete Institute, 2001.

Momber, A.W., Schulz, R.-R. *Handbuch der Oberflächenbearbeitung Beton*. Basel: Birkhäuser, 2006.

Müller, H.S., Fenchel, M. *Zerstörungsfreie Ortung von Gefügestörungen im Beton*, pp. 315–334. Stuttgart: Fraunhofer-Informationszentrum Raum und Bau, 2006.

Page, C.L., Havdahl, J. Electrochemical Monitoring of Corrosion of Steel in Microsilica Cement Pastes. *Materiaux et Constructions* 18 (1985), no. 103, s. 41–47.

Polder, R.B., Peelen, W.H.A., Raupach, M., Reichling, K. Economic Effects of Full Corrosion Surveys for Aging Concrete Structures. In *Materials and Corrosion* 64 (2013), Nr. 2, S. 105–110 ISSN 1521–4176.

Raupach, M. Zur chloridinduzierten Makroelementkorrosion von Stahl in Beton. *Schriftenreihe des deutschen Ausschusses für Stahlbeton* (1992), no. 433, PhD thesis.

Raupach, M. Durability of Marine and Coastal Structures—The High Quality Cover and Monitoring Approach. Dundee, Scotland, 2005. In *Proceedings of the 6th International Congress Global Construction: Ultimate Concrete Opportunities*, 5–7 July 2005, Dundee, Scotland, 10 pages.

Raupach, M. Patch Repair on Reinforced Concrete Structures—Model Investigations on the Required Size and Practical Consequences. *Cement and Concrete Composites* 28 (2006), no. 8, pp. 679–684.

Raupach, M., Büttner, T., Maintz, H. Retrofitting AACHEN Cathedral with a Flexible Textile Reinforced Mortar Bandage. In *Proceedings of the Third International Workshop on Civil Structural Health Monitoring: Conservation of Heritage Structures Using FRM and SHM*, ed. N. Banthia and A. Mufti, Ottawa, Canada, August 11–13, 2010, pp. 47–60. Vancouver: University of British Columbia, 2010.

Raupach, M., Dauberschmidt, C., Wolff, L., Harnisch, J. Monitoring der Feuchteverteilung in Beton: Monitoring the Humidity Distribution in Concrete. In *Beton* 57 (2007), Nr. 1+2, S. 20–26.

Raupach, M., Gulikers, J., Reichling, K. Condition Survey with Embedded Sensors Regarding Reinforcement Corrosion. In *Materials and Corrosion* 64 (2013), Nr. 2, S. 141–146 ISSN 1521–4176.

Reichling, K., Raupach, M., Broomfield, J., Gulikers, J., L'Hostis, V., Keßler, S., Osterminski, K., Pepenar, I., Schneck, U., Sergi, G., Tache, G. Full Surface Inspection Methods Regarding Reinforcement Corrosion of Concrete Structures. In *Materials and Corrosion* 64 (2013), Nr. 2, S. 116–127 ISSN 1521–4176.

Raupach, M., Reichling, K., Broomfield, J., Gulikers, J., Schneck, U., Serdar, M., Pepenar, I. Inspection Strategies for Reinforcement Corrosion Surveys. In *Materials and Corrosion* 64 (2013), Nr. 2, S. 111–115 ISSN 1521–4176.

Raupach, M., Wolff, L. Reduktion der Bewehrungsüberdeckung bei vorhandener Beschichtung bei Parkhaus-Neubauten. *Kurzberichte aus der Bauforschung* 46 (2005), no. 2, pp. 82–88.

Reichling, K., Weichold, O., Raupach, M. What Do We Learn About the Electrical Resistivity of Concrete? *Restoration of Buildings and Monuments* 18 (2012), no. 5, s. 265–274.

RILEM TC 154-EMC; Elsner, B., Andrade, C., Gulikers, J., Raupach, M. Half-Cell Potential Measurements—Potential Mapping on Reinforced Concrete Structures. *Materials and Structures* (RILEM) 36 (2003), no. 261, pp. 461–471.

Sasse, H.R., Schäfer, H.G., Bäätjer, G., Borg, G., et al. Verstärken von Betonbauteilen—Sachstandsbericht. *Schriftenreihe des Deutschen Ausschusses für Stahlbeton* (1996), no. 467.

Schießl, P., Raupach, M. Laboruntersuchungen und Berechnungen zum Einfluß der Rißbreite auf die chloridinduzierte Korrosion von Stahl in Beton. *Bauingenieur* 69 (1994), no. 11, pp. 439–445.

Schneck, U. Electrochemical Chloride Extraction. In *Concrete Repair: A Practical Guide*, ed. M.G. Grantham, pp. 147–168. London: Routledge, Taylor & Francis Group, 2011.

Schulz, R.-R. *Beton als Beschichtungsuntergrund: Über die Prüfung des Festigkeitsverhaltens von Betonoberflächen mit dem Abreißversuch*. Dissertation, Technische Hochschule, Aachen, Fachbereich 3, 1984.

Tritthart, J., Baumgartner, E. Umverteilung der Inhaltsstoffe der Porenlösung von Beton nach unterschiedlichen Korrosionsschutzmaßnahmen. In *Verstärken und Instandsetzen von Betonkonstruktionen 2007, Berichtsband der 6. Intern*, ed. W. Kusterle et al., Fachtagung, Innsbruck, January 25–26, 2007, pp. 137–144. Innsbruck: Institut für Konstruktion und Materialwissenschaften, 2007.

Tuutti, K. Corrosion of Steel in Concrete. *CBI Research* (1982), no. fo. 4:82.

Warkus, J. Bestimmung des Wassergehaltes von Beton und Naturstein durch Messung des Elektrolytwiderstandes—Grundlagen. In *Zustandserfassung von Bauwerken—Neue Verfahren und System, 29*, Aachener Baustofftag, Aachen, November 18, 2003. Aachen: Institut für Bauforschung, 2003.

Warkus, J. Einfluss der Bauteilgeometrie auf die Korrosionsgeschwindigkeit von Stahl in Beton bei Markoelementbildung. Dissertation, Technische Hochschule, Aachen, Fachbereich 3, 2012.

Wolff, L., Bruns, M., Raupach, M. Influence of Cracks on the Corrosion Risk at Parking Decks—Theory and Practice. In *3. Kolloquium Erhaltung von Bauwerken*, ed. M. Raupach, Esslingen, January 22–23, 2013, pp. 211–221. Ostfildern: Technische Akademie Esslingen, 2013.

Wolff, L., Raupach, M. Beschichtungsschäden-Schadensmechanismen und Lösungsansätze: Coating Damages, Damage Mechanisms and proposed Solutions. Ostfildern: Technische Akademie Esslingen, 2008. In *Verkehrsbauten Schwerpunkt Parkhäuser/Brücken*. 3. Kolloquium, Ostfildern, 29. und 30. Januar 2008, (Gieler-Breßmer, S. (Ed.)), S. 313–324.

EN 1504 norm series

(DIN) EN 1504-1: 2005. *Products and Systems for the Protection and Repair of Concrete Structures—Definitions, Requirements, Quality Control and Evaluation of Conformity—Part 1: Definitions.* German version, EN 1504-1:2005.

(DIN) EN 1504-2:2005. *Products and Systems for the Protection and Repair of Concrete Structures—Definitions, Requirements, Quality Control and Evaluation of Conformity—Part 2: Surface Protection Systems for Concrete.* German version, EN 1504-2:2004.

(DIN) EN 1504-3:2006. *Products and Systems for the Protection and Repair of Concrete Structures—Definitions, Requirements, Quality Control and Evaluation of Conformity—Part 3: Structural and Non-structural Repair.* German version, EN 1504-3:2005.

(DIN) EN 1504-4:2005. *Products and Systems for the Protection and Repair of Concrete Structures—Definitions, Requirements, Quality Control and Evaluation of Conformity—Part 4: Structural Bonding.* German version, EN 1504-4:2004.

(DIN) EN 1504-5:2013. *Products and Systems for the Protection and Repair of Concrete Structures—Definitions, Requirements, Quality Control and Evaluation of Conformity—Part 5: Concrete Injection.* German version, EN 1504-5:2013.

(DIN) EN 1504-6:2006. *Products and Systems for the Protection and Repair of Concrete Structures—Definitions, Requirements, Quality Control and Evaluation of Conformity—Part 6: Anchoring of Reinforcing Steel Bar.* German version, EN 1504-6:2006.

(DIN) EN 1504-7:2006. *Products and Systems for the Protection and Repair of Concrete Structures—Definitions, Requirements, Quality Control and Evaluation of Conformity—Part 7: Reinforcement Corrosion Protection.* German version, EN 1504-7:2006.

(DIN) EN 1504-8:2005. *Products and Systems for the Protection and Repair of Concrete Structures—Definitions, Requirements, Quality Control and Evaluation of Conformity—Part 8: Quality Control and Evaluation of Conformity.* German version, EN 1504-8:2004.

(DIN) EN 1504-9:2008. *Products and Systems for the Protection and Repair of Concrete Structures—Definitions, Requirements, Quality Control and Evaluation of Conformity—Part 9: General Principles for the Use of Products and Systems.* German version, EN 1504-9:2008.

(DIN) EN 1504-10:2004. *Products and Systems for the Protection and Repair of Concrete Structures—Definitions, Requirements, Quality Control and Evaluation of Conformity—Part 10: Site Application of Products and Systems and Quality Control of the Works.* German version, EN 1504-10:2003.

Index

Printed and bound by CPI Group (UK) Ltd, Croydon, CR0 4YY

01/11/2024

01782605-0010